Sensor Network Protocols

Sensor Network Protocols

IMAD MAHGOUB

Florida Atlantic University

Boca Raton, Florida, U.S.A.

MOHAMMAD ILYAS

Florida Atlantic University

Boca Raton, Florida, U.S.A.

Taylor & Francis
Taylor & Francis Group
Boca Raton   London   New York

A CRC title, part of the Taylor & Francis imprint, a member of the
Taylor & Francis Group, the academic division of T&F Informa plc.

The material was previously published in *Handbook of Sensor Networks: Compact Wireless and Wired Sensing Systems.*
© CRC Press LLC 2005.

Published in 2006 by
CRC Press
Taylor & Francis Group
6000 Broken Sound Parkway NW, Suite 300
Boca Raton, FL 33487-2742

International Standard Book Number-10: 0-8493-7036-1 (Hardcover)
International Standard Book Number-13: 978-0-8493-7036-6 (Hardcover)
Library of Congress Card Number 2005053837

Library of Congress Cataloging-in-Publication Data

Sensor network protocols / [edited by] Imad Mahgoub, Mohammad Ilyas.
    p. cm.
  Includes bibliographical references and index.
  ISBN 0-8493-7036-1 (alk. paper)
   1. Sensor networks. 2. Computer network protocols. I. Mahgoub, Imad. II. Ilyas, Mohammad, 1953-.

TK7872.D48S44 2005
681'.2--dc22   2005053837

Taylor & Francis Group
is the Academic Division of Informa plc.

Visit the Taylor & Francis Web site at
http://www.taylorandfrancis.com

and the CRC Press Web site at
http://www.crcpress.com

Preface

Advances in wireless communications and microelectronic mechanical systems technologies have enabled the development of networks of a large number of small inexpensive, low-power multifunctional sensors. These wireless sensor networks present a very interesting and challenging area and have tremendous potential applications. Communication protocols are the heart and soul of any communication network and the same is true for the sensor networks. This book deals with wireless sensor network protocols.

Wireless sensor networks consist of a large number of sensor nodes that may be randomly and densely deployed. Sensor nodes are small electronic components capable of sensing many types of information from the environment including temperature, light, humidity, radiation, the presence or nature of biological organisms, geological features, seismic vibrations, specific types of computer data, and more. Recent advancements have made it possible to make these components small, powerful, and energy efficient and they can now be manufactured cost-effectively in quantity for specialized telecommunications applications. The sensor nodes are very small in size and are capable of gathering, processing, and communicating information to other nodes and to the outside world.

This book is expected to capture the current state of protocols for sensor networks. The book has a total of eleven articles written by experts from around the world. These articles were previously published in the *Handbook of Sensor Networks: Compact Wireless and Wired Sensing Systems* by CRC Press, 2005.

The targeted audience for the book includes professionals who are designers and/or planners for emerging telecommunication networks, researchers (faculty members and graduate students), and those who would like to learn about this field.

Although the book is not precisely a textbook, it can certainly be used as a textbook for graduate courses and research-oriented courses that deal with wireless sensor networks. Any comments from the readers will be highly appreciated.

Many people have contributed to this book in their unique ways. The first and the foremost group that deserves immense gratitude is the group of highly talented and skilled researchers who have contributed eleven articles to this book. All of them have been extremely cooperative and professional. It has also been a pleasure to work with Ms. Nora Konopka, Ms. Helena Redshaw, and Ms. Allison Taub of Taylor & Francis/CRC Press and we are extremely gratified for their support and professionalism. Our families have extended their unconditional love and strong support throughout this project and they all deserve a very special thanks.

Imad Mahgoub and Mohammad Ilyas
Boca Raton, Florida

Editors

Imad Mahgoub, Ph.D., received his B.Sc. degree in electrical engineering from the University of Khartoum, Khartoum, Sudan, in 1978. From 1978 to 1981, he worked for the Sudan Shipping Line Company, Port Sudan, Sudan, as an electrical and electronics engineer. He received his M.S. in applied mathematics in 1983 and his M.S. in electrical and computer engineering in 1986, both from North Carolina State University. In 1989, he received his Ph.D. in computer engineering from The Pennsylvania State University.

Since August 1989, Dr. Mahgoub has been with the College of Engineering at Florida Atlantic University, Boca Raton, Florida, where he is currently professor of computer science and engineering. He is the director of the Computer Science and Engineering Department Mobile Computing Laboratory at Florida Atlantic University.

Dr. Mahgoub has conducted successful research in various areas, including mobile computing; interconnection networks; performance evaluation of computer systems; and advanced computer architecture. He has published over 80 research articles and supervised three Ph.D. dissertations and 22 M.S. theses to completion. He has served as a consultant to industry. Dr. Mahgoub served as a member of the executive committee/program committee of the 1998, 1999, and 2000 IEEE International Performance, Computing and Communications Conferences. He has served on the program committees of several international conferences and symposia. He was the vice chair of the 2003, 2004, and 2005 International Symposium on Performance Evaluation of Computer and Telecommunication Systems. Dr. Mahgoub is a senior member of IEEE and a member of ACM.

Mohammad Ilyas, Ph.D., received his B.Sc. degree in electrical engineering from the University of Engineering and Technology, Lahore, Pakistan, in 1976. From March 1977 to September 1978, he worked for the Water and Power Development Authority in Pakistan. In 1978, he was awarded a scholarship for his graduate studies and he completed his M.S. degree in electrical and electronic engineering in June 1980 at Shiraz University, Shiraz, Iran. In September 1980, he joined the doctoral program at Queen's University in Kingston, Ontario, Canada; he completed his Ph.D. degree in 1983. Dr. Ilyas' doctoral research was about switching and flow control techniques in computer communication networks. Since September 1983, he has been with the College of Engineering at Florida Atlantic University, Boca Raton, Florida, where he is currently associate dean for graduate studies and research. From 1994 to 2000, he was chair of the department. During the 1993–1994 academic year, he was on his sabbatical leave with the Department of Computer Engineering, King Saud University, Riyadh, Saudi Arabia.

Dr. Ilyas has conducted successful research in various areas, including traffic management and congestion control in broadband/high-speed communication networks; traffic characterization; wireless communication networks; performance modeling; and simulation. He has published one book, three handbooks, and over 140 research articles. He has supervised 10 Ph.D. dissertations and more than 35 M.S. theses to completion. Dr. Ilyas has been a consultant to several national and international organizations; a senior member of IEEE, he is an active participant in several IEEE technical committees and activities.

Contributors

Özgür B. Akan
Georgia Institute of
  Technology
Atlanta, Georgia

Jamal N. Al-Karaki
Iowa State University
Ames, Iowa

Jacir L. Bordim
ATR — Adaptive
  Communications Research
  Laboratories
Kyoto, Japan

Erdal Cayirci
Istanbul Technical University
Istanbul, Turkey

Krishnendu Chakrabarty
Duke University
Durham, North Carolina

**Duminda
  Dewasurendra**
Virginia Polytechnic Institute
  and State University
Blacksburg, Virginia

Jessica Feng
University of California at Los
  Angeles
Los Angeles, California

Joel I. Goodman
MIT Lincoln Laboratory
Lexington, Massachusetts

Martin Haenggi
University of Notre Dame
Notre Dame, Indiana

Hossam Hassanein
Queen's University
Kingston, Ontario, Canada

S. Sitharama Iyengar
Louisiana State University
Baton Rouge, Louisiana

Ram Kalidindi
Louisiana State University
Baton Rouge, Louisiana

Ahmed E. Kamal
Iowa State University
Ames, Iowa

Rajgopal Kannan
Louisiana State University
Baton Rouge, Louisiana

Farinaz Koushanfar
University of California at
  Berkeley
Berkeley, California

David R. Martinez
MIT Lincoln Laboratory
Lexington, Massachusetts

Amitabh Mishra
Virginia Polytechnic Institute
  and State University
Blacksburg, Virginia

Koji Nakano
Hiroshima University
Higashi-Hiroshima, Japan

Dragan Petrovic
University of California at
  Berkeley
Berkeley, California

Miodrag Potkonjak
University of California at
  Los Angeles
Los Angeles, California

Jan M. Rabaey
University of California at
  Berkeley
Berkeley, California

Lydia Ray
Louisiana State University
Baton Rouge, Louisiana

Albert I. Reuther
MIT Lincoln Laboratory
Lexington, Massachusetts

Rahul C. Shah
University of California at
  Berkeley
Berkeley, California

Weilian Su
Georgia Institute of
  Technology
Atlanta, Georgia

Vishnu Swaminathan
Duke University
Durham, North Carolina

Quanhong Wang
Queen's University
Kingston, Ontario, Canada

Yi Zou
Duke University
Durham, North Carolina

Contents

1

Opportunities and Challenges in Wireless Sensor Networks

Martin Haenggi

University of Notre Dame

1.1 Introduction

Due to advances in wireless communications and electronics over the last few years, the development of networks of low-cost, low-power, multifunctional sensors has received increasing attention. These sensors are small in size and able to sense, process data, and communicate with each other, typically over an RF (radio frequency) channel. A sensor network is designed to detect events or phenomena, collect and process data, and transmit sensed information to interested users. Basic features of sensor networks are:

- Self-organizing capabilities
- Short-range broadcast communication and multihop routing
- Dense deployment and cooperative effort of sensor nodes
- Frequently changing topology due to fading and node failures
- Limitations in energy, transmit power, memory, and computing power

These characteristics, particularly the last three, make sensor networks different from other wireless ad hoc or mesh networks.

Clearly, the idea of mesh networking is not new; it has been suggested for some time for wireless Internet access or voice communication. Similarly, small computers and sensors are not innovative per se. However, combining small sensors, low-power computers, and radios makes for a new technological platform that has numerous important uses and applications, as will be discussed in the next section.

1.2 Opportunities

1.2.1 Growing Research and Commercial Interest

Research and commercial interest in the area of wireless sensor networks are currently growing exponentially, which is manifested in many ways:

- The number of Web pages (Google: 26,000 hits for sensor networks; 8000 for wireless sensor networks in August 2003)
- The increasing number of
  - Dedicated annual workshops, such as IPSN (information processing in sensor networks); SenSys; EWSN (European workshop on wireless sensor networks); SNPA (sensor network protocols and applications); and WSNA (wireless sensor networks and applications)
  - Conference sessions on sensor networks in the communications and mobile computing communities (ISIT, ICC, Globecom, INFOCOM, VTC, MobiCom, MobiHoc)
  - Research projects funded by NSF (apart from ongoing programs, a new specific effort now focuses on sensors and sensor networks) and DARPA through its SensIT (sensor information technology), NEST (networked embedded software technology), MSET (multisensor exploitation), UGS (unattended ground sensors), NETEX (networking in extreme environments), ISP (integrated sensing and processing), and communicator programs

Special issues and sections in renowned journals are common, e.g., in the *IEEE Proceedings* [1] and signal processing, communications, and networking magazines. Commercial interest is reflected in investments by established companies as well as start-ups that offer general and specific hardware and software solutions.

Compared to the use of a few expensive (but highly accurate) sensors, the strategy of deploying a large number of inexpensive sensors has significant advantages, at smaller or comparable total system cost: much higher spatial resolution; higher robustness against failures through distributed operation; uniform coverage; small obtrusiveness; ease of deployment; reduced energy consumption; and, consequently, increased system lifetime. The main point is to position sensors close to the source of a potential problem phenomenon, where the acquired data are likely to have the greatest benefit or impact.

Pure sensing in a fine-grained manner may revolutionize the way in which complex physical systems are understood. The addition of actuators, however, opens a completely new dimension by permitting management and manipulation of the environment at a scale that offers enormous opportunities for almost every scientific discipline. Indeed, Business 2.0 (http://www.business2.com/) lists sensor robots as one of "six technologies that will change the world," and *Technology Review* at MIT and Globalfuture identify WSNs as one of the "10 emerging technologies that will change the world" (http://www.globalfuture.com/mit-trends2003.htm). The combination of sensor network technology with MEMS and nanotechnology will greatly reduce the size of the nodes and enhance the capabilities of the network.

The remainder of this chapter lists and briefly describes a number of applications for wireless sensor networks, grouped into different categories. However, because the number of areas of application is growing rapidly, every attempt at compiling an exhaustive list is bound to fail.

1.2.2 Applications

1.2.2.1 General Engineering

- *Automotive telematics.* Cars, which comprise a network of dozens of sensors and actuators, are networked into a system of systems to improve the safety and efficiency of traffic.
- *Fingertip accelerometer virtual keyboards.* These devices may replace the conventional input devices for PCs and musical instruments.
- *Sensing and maintenance in industrial plants.* Complex industrial robots are equipped with up to 200 sensors that are usually connected by cables to a main computer. Because cables are expensive

and subject to wear and tear caused by the robot's movement, companies are replacing them by wireless connections. By mounting small coils on the sensor nodes, the principle of induction is exploited to solve the power supply problem.

- *Aircraft drag reduction.* Engineers can achieve this by combining flow sensors and blowing/sucking actuators mounted on the wings of an airplane.
- *Smart office spaces.* Areas are equipped with light, temperature, and movement sensors, microphones for voice activation, and pressure sensors in chairs. Air flow and temperature can be regulated locally for one room rather than centrally.
- *Tracking of goods in retail stores.* Tagging facilitates the store and warehouse management.
- *Tracking of containers and boxes.* Shipping companies are assisted in keeping track of their goods, at least until they move out of range of other goods.
- *Social studies.* Equipping human beings with sensor nodes permits interesting studies of human interaction and social behavior.
- Commercial and residential security.

1.2.2.2 Agriculture and Environmental Monitoring

- *Precision agriculture.* Crop and livestock management and precise control of fertilizer concentrations are possible.
- *Planetary exploration.* Exploration and surveillance in inhospitable environments such as remote geographic regions or toxic locations can take place.
- *Geophysical monitoring.* Seismic activity can be detected at a much finer scale using a network of sensors equipped with accelerometers.
- *Monitoring of freshwater quality.* The field of hydrochemistry has a compelling need for sensor networks because of the complex spatiotemporal variability in hydrologic, chemical, and ecological parameters and the difficulty of labor-intensive sampling, particularly in remote locations or under adverse conditions. In addition, buoys along the coast could alert surfers, swimmers, and fishermen to dangerous levels of bacteria.
- *Zebranet.* The Zebranet project at Princeton aims at tracking the movement of zebras in Africa.
- *Habitat monitoring.* Researchers at UC Berkeley and the College of the Atlantic in Bar Harbor deployed sensors on Great Duck Island in Maine to measure humidity, pressure, temperature, infrared radiation, total solar radiation, and photosynthetically active radiation (see http://www.greatduckisland.net/).
- *Disaster detection.* Forest fire and floods can be detected early and causes can be localized precisely by densely deployed sensor networks.
- *Contaminant transport.* The assessment of exposure levels requires high spatial and temporal sampling rates, which can be provided by WSNs.

1.2.2.3 Civil Engineering

- *Monitoring of structures.* Sensors will be placed in bridges to detect and warn of structural weakness and in water reservoirs to spot hazardous materials. The reaction of tall buildings to wind and earthquakes can be studied and material fatigue can be monitored closely.
- *Urban planning.* Urban planners will track groundwater patterns and how much carbon dioxide cities are expelling, enabling them to make better land-use decisions.
- *Disaster recovery.* Buildings razed by an earthquake may be infiltrated with sensor robots to locate signs of life.

1.2.2.4 Military Applications

- *Asset monitoring and management.* Commanders can monitor the status and locations of troops, weapons, and supplies to improve military command, control, communications, and computing (C4).

- *Surveillance and battle-space monitoring.* Vibration and magnetic sensors can report vehicle and personnel movement, permitting close surveillance of opposing forces.
- *Urban warfare.* Sensors are deployed in buildings that have been cleared to prevent reoccupation; movements of friend and foe are displayed in PDA-like devices carried by soldiers. Snipers can be localized by the collaborative effort of multiple acoustic sensors.
- *Protection.* Sensitive objects such as atomic plants, bridges, retaining walls, oil and gas pipelines, communication towers, ammunition depots, and military headquarters can be protected by intelligent sensor fields able to discriminate between different classes of intruders. Biological and chemical attacks can be detected early or even prevented by a sensor network acting as a warning system.
- *Self-healing minefields.* The self-healing minefield system is designed to achieve an increased resistance to dismounted and mounted breaching by adding a novel dimension to the minefield. Instead of a static complex obstacle, the self-healing minefield is an intelligent, dynamic obstacle that senses relative positions and responds to an enemy's breaching attempt by physical reorganization.

1.2.2.5 Health Monitoring and Surgery

- *Medical sensing.* Physiological data such as body temperature, blood pressure, and pulse are sensed and automatically transmitted to a computer or physician, where it can be used for health status monitoring and medical exploration. Wireless sensing bandages may warn of infection. Tiny sensors in the blood stream, possibly powered by a weak external electromagnetic field, can continuously analyze the blood and prevent coagulation and thrombosis.
- *Micro-surgery.* A swarm of MEMS-based robots may collaborate to perform microscopic and minimally invasive surgery.

The opportunities for wireless sensor networks are ubiquitous. However, a number of formidable challenges must be solved before these exciting applications may become reality.

1.3 Technical Challenges

Populating the world with networks of sensors requires a fundamental understanding of techniques for connecting and managing sensor nodes with a communication network in scalable and resource-efficient ways. Clearly, sensor networks belong to the class of ad hoc networks, but they have specific characteristics that are not present in general ad hoc networks.

Ad hoc and sensor networks share a number of challenges such as energy constraints and routing. On the other hand, general ad hoc networks most likely induce traffic patterns different from sensor networks, have other lifetime requirements, and are often considered to consist of *mobile* nodes [2–4]. In WSNs, most nodes are static; however, the network of basic sensor nodes may be overlaid by more powerful mobile sensors (robots) that, guided by the basic sensors, can move to interesting areas or even track intruders in the case of military applications.

Network nodes are equipped with wireless transmitters and receivers using antennas that may be omnidirectional (isotropic radiation), highly directional (point-to-point), possibly steerable, or some combination thereof. At a given point in time, depending on the nodes' positions and their transmitter and receiver coverage patterns, transmission power levels, and cochannel interference levels, a wireless connectivity exists in the form of a random, multihop graph between the nodes. This ad hoc topology may change with time as the nodes move or adjust their transmission and reception parameters.

Because the most challenging issue in sensor networks is *limited and unrechargeable* energy provision, many research efforts aim at improving the energy efficiency from different aspects. In sensor networks, energy is consumed mainly for three purposes: *data transmission, signal processing,* and *hardware operation* [5]. It is desirable to develop energy-efficient processing techniques that minimize power requirements across all levels of the protocol stack and, at the same time, minimize message passing for network control and coordination.

1.3.1 Performance Metrics

To discuss the issues in more detail, it is necessary to examine a list of metrics that determine the performance of a sensor network:

- *Energy efficiency/system lifetime.* The sensors are battery operated, rendering energy a very scarce resource that must be wisely managed in order to extend the lifetime of the network [6].
- *Latency.* Many sensor applications require delay-guaranteed service. Protocols must ensure that sensed data will be delivered to the user within a certain delay. Prominent examples in this class of networks are certainly the sensor-actuator networks.
- *Accuracy.* Obtaining accurate information is the primary objective; accuracy can be improved through joint detection and estimation. Rate distortion theory is a possible tool to assess accuracy.
- *Fault tolerance.* Robustness to sensor and link failures must be achieved through redundancy and collaborative processing and communication.
- *Scalability.* Because a sensor network may contain thousands of nodes, scalability is a critical factor that guarantees that the network performance does not significantly degrade as the network size (or node density) increases.
- *Transport capacity/throughput.* Because most sensor data must be delivered to a single base station or fusion center, a *critical area* in the sensor network exists (the gray area in Figure 1.1.), whose sensor nodes must relay the data generated by virtually all nodes in the network. Thus, the traffic load at those critical nodes is heavy, even when the average traffic rate is low. Apparently, this area has a paramount influence on system lifetime, packet end-to-end delay, and scalability.

Because of the interdependence of energy consumption, delay, and throughput, all these issues and metrics are tightly coupled. Thus, the design of a WSN necessarily consists of the resolution of numerous trade-offs, which also reflects in the network protocol stack, in which a cross-layer approach is needed instead of the traditional layer-by-layer protocol design.

1.3.2 Power Supply

The most difficult constraints in the design of WSNs are those regarding the minimum energy consumption necessary to drive the circuits and possible microelectromechanical devices (MEMS) [5, 7, 8]. The energy problem is aggravated if actuators are present that may be substantially hungrier for power than the sensors. When miniaturizing the node, the energy density of the power supply is the primary issue. Current technology yields batteries with approximately 1 J/mm³ of energy, while capacitors can achieve as much as 1 mJ/mm³. If a node is designed to have a relatively short lifespan, for example, a few months, a battery is a logical solution. However, for nodes that can generate sensor readings for long periods of time, a charging

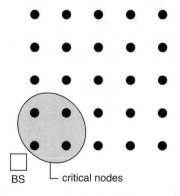

FIGURE 1.1 Sensor network with base station (or fusion center). The gray-shaded area indicates the critical area whose nodes must relay all the packets.

method for the supply is preferable. Currently, research groups are investigating the use of solar cells to charge capacitors with photocurrents from the ambient light sources. Solar flux can yield power densities of approximately 1 mW/mm^2. The energy efficiency of a solar cell ranges from 10 to 30% in current technologies, giving 300 μW in full sunlight in the best-case scenario for a 1-mm^2 solar cell operating at 1 V. Series-stacked solar cells will need to be utilized in order to provide appropriate voltages.

Sensor acquisition can be achieved at 1 nJ per sample, and modern processors can perform computations as low as 1 nJ per instruction. For wireless communications, the primary candidate technologies are based on RF and optical transmission techniques, each of which has its advantages and disadvantages. RF presents a problem because the nodes may offer very limited space for antennas, thereby demanding very short-wavelength (i.e., high-frequency) transmission, which suffers from high attenuation. Thus, communication in that regime is not currently compatible with low-power operation. Current RF transmission techniques (e.g., Bluetooth [9]) consume about 100 nJ per bit for a distance of 10 to 100 m, making communication very expensive compared to acquisition and processing.

An alternative is to employ free-space optical transmission. If a line-of-sight path is available, a well-designed free-space optical link requires significantly lower energy than its RF counterpart, currently about 1 nJ per bit. The reason for this power advantage is that optical transceivers require only simple baseband analog and digital circuitry and no modulators, active filters, and demodulators. Furthermore, the extremely short wavelength of visible light makes it possible for a millimeter-scale device to emit a narrow beam, corresponding to an antenna gain of roughly five to six orders of magnitude compared to an isotropic radiator. However, a major disadvantage is that the beam needs to be pointed very precisely at the receiver, which may be prohibitively difficult to achieve.

In WSNs, where sensor sampling, processing, data transmission, and, possibly, actuation are involved, the trade-off between these tasks plays an important role in power usage. Balancing these parameters will be the focus of the design process of WSNs.

1.3.3 Design of Energy-Efficient Protocols

It is well acknowledged that *clustering* is an efficient way to save energy for static sensor networks [10–13]. Clustering has three significant differences from conventional clustering schemes. First, data compression in the form of distributed source coding is applied within a cluster to reduce the number of packets to be transmitted [14, 15]. Second, the *data-centric* property makes an identity (e.g., an address) for a sensor node obsolete. In fact, the user is often interested in phenomena occurring in a specified area [16], rather than in an individual sensor node. Third, randomized rotation of cluster heads helps ensure a balanced energy consumption [11].

Another strategy to increase energy efficiency is to use *broadcast and multicast trees* [6, 17, 18], which take advantage of the *broadcast property* of omnidirectional antennas. The disadvantage is that the high computational complexity may offset the achievable benefit. For sensor networks, this *one-to-many* communication scheme is less important; however, because all data must be delivered to a single destination, the traffic scheme (for application traffic) is the opposite, i.e., *many to one*. In this case, clearly the *wireless multicast advantage* offers less benefit, unless path diversity or cooperative diversity schemes are implemented [19, 20].

The exploitation of *sleep modes* [21, 22] is imperative to prevent sensor nodes from wasting energy in receiving packets unintended for them. Combined with efficient medium access protocols, the "sleeping" approach could reach optimal energy efficiency without degradation in throughput (but at some penalty in delay).

1.3.4 Capacity/Throughput

Two parameters describe the network's capability to carry traffic: *transport capacity* and *throughput*. The former is a distance-weighted sum capacity that permits evaluation of network performance. Throughput is a traditional measure of how much traffic can be delivered by the network [23–30]. In a packet network, the (network-layer) throughput may be defined as the expected number of successful packet transmissions of a given node per timeslot.

The capacity of wireless networks in general is an active area of research in the information theory community. The results obtained mostly take the form of scaling laws or "order-of" results; the prefactors are difficult to determine analytically. Important results include the scaling law for point-to-point coding, which shows that the throughput decreases with $1/\sqrt{N}$ for a network with N nodes [23]. Newer results [28] permit network coding, which yields a slightly more optimistic scaling behavior, although at high complexity. Grossglauser and Tse [26] have shown that mobility may keep the per-node capacity constant as the network grows, but that benefit comes at the cost of unbounded delay.

The throughput is related to (error-free) transmission rate of each transmitter, which, in turn, is upper bounded by the channel capacity. From the pure information theoretic point of view, the capacity is computed based on the ergodic channel assumption, i.e., the code words are long compared to the coherence time of the channel. This Shannon-type capacity is also called *throughput capacity* [31]. However, in practical networks, particularly with delay-constrained applications, this capacity cannot provide a helpful indication of the channel's ability to transmit with a small probability of error.

Moreover, in the multiple-access system, the corresponding power allocation strategies for maximum achievable capacity always favor the "good" channels, thus leading to unfairness among the nodes. Therefore, for delay-constrained applications, the channel is usually assumed to be nonergodic and the capacity is a random variable, instead of a constant in the classical definition by Shannon. For a delay-bound D, the channel is often assumed to be block fading with block length D, and a *composite channel* model is appropriate when specifying the capacity. Correspondingly, given the noise power, the channel state (a random variable in the case of fading channels), and power allocation, new definitions for *delay-constrained* systems have been proposed [32–35].

1.3.5 Routing

In ad hoc networks, routing protocols are expected to implement three main functions: *determining and detecting network topology changes* (e.g., breakdown of nodes and link failures); *maintaining network connectivity*; and *calculating and finding proper routes*. In sensor networks, up-to-date, less effort has been given to routing protocols, even though it is clear that ad hoc routing protocols (such as *destination-sequenced distance vector* (DSDV), *temporally-ordered routing algorithm* (TORA), *dynamic source routing* (DSR), and *ad hoc on-demand distance vector* (AODV) [4, 36–39]) are not suited well for sensor networks since the main type of traffic in WSNs is "many to one" because all nodes typically report to a single base station or fusion center. Nonetheless, some merits of these protocols relate to the features of sensor networks, like *multihop communication* and *QoS routing* [39]. Routing may be associated with data compression [15] to enhance the scalability of the network.

1.3.6 Channel Access and Scheduling

In WSNs, scheduling must be studied at two levels: the *system level* and the *node level*. At the node level, a scheduler determines which flow among all multiplexing flows will be eligible to transmit next (the same concept as in traditional wired scheduling); at the system level, a scheme determines which nodes will be transmitting. System-level scheduling is essentially a medium access (MAC) problem, with the goal of minimum collisions and maximum spatial reuse — a topic receiving great attention from the research community because it is tightly coupled with energy efficiency and throughput.

Most of the current wireless scheduling algorithms aim at improved *fairness, delay, robustness* (with respect to network topology changes) and *energy efficiency* [62, 64, 65, 66]. Some also propose a distributed implementation, in contrast to the centralized implementation in wired or cellular networks, which originated from general fair queuing. Also, wireless (or sensor) counterparts of other wired scheduling classes, like *priority scheduling* [67, 68] and *earliest deadline first (EDF)* [69], confirm that prioritization is necessary to achieve *delay balancing* and *energy balancing*.

The main problem in WSNs is that all the sensor data must be forwarded to a base station via multihop routing. Consequently, the traffic pattern is highly nonuniform, putting a high burden on the sensor nodes close to the base station (the critical nodes in Figure 1.1). The scheduling algorithm and routing protocols must aim at *energy and delay balancing*, ensuring that packets originating close and far away from the base station experience a comparable delay, and that the critical nodes do not die prematurely due to the heavy relay traffic [40].

At this point, due to the complexity of scheduling algorithms and the wireless environment, most performance measures are given through simulation rather than analytically. Moreover, *medium access* and *scheduling* are usually considered separately. When discussing scheduling, the system is assumed to have a single user; whereas in the MAC layer, all flows multiplexing at the node are treated in the same way, i.e., a default FIFO buffer is assumed to schedule flows. It is necessary to consider them jointly to optimize performance figures such as delay, throughput, and packet loss probability.

Because of the bursty nature of the network traffic, random access methods are commonly employed in WSNs, with or without carrier sense mechanisms. For illustrative purposes, consider the simplest sensible MAC scheme possible: all nodes are transmitting packets independently in every timeslot with the same transmit probability p at equal transmitting power levels; the next-hop receiver of every packet is one of its neighbors. The packets are of equal length and fit into one timeslot. This MAC scheme was considered in Silvester and Kleinrock [41], Hu [42], and Haenggi [43]. The resulting (per-node) throughput turns out to be a polynomial in p of order N, where N is the number of nodes in the network.

A typical throughput polynomial is shown in Figure 1.2. At $p = 0$, the derivative is 1, indicating that, for small p, the throughput equals p. This is intuitive because there are few collisions for small p and the throughput $g(p)$ is approximately linear. The region in which the packet loss probability is less than 10% can be denoted as the *collisionless* region. It ranges from 0 to about $p_{max}/8$. The next region, up to p_{max}, is the practical region in which energy consumption (transmission attempts) is traded off against throughput; it is therefore called the *trade-off* region. The difference $p - g(p)$ is the *interference loss*. For small networks, all N nodes interfere with each other because spatial reuse is not possible: If more than one node is transmitting, a collision occurs and all packets are lost. Thus, the (per-node) throughput is $p(1 - p)^{N-1}$, and the optimum transmit probability is $1/N$. The maximum throughput is $(1 - 1/N)^{N-1}/N$. With increasing N, the throughput approaches $1/(eN)$, as pointed out in Silvester and Kleinrock [41] and LaMaire et al. [44]. Therefore the difference $p_{max} - 1/N$ is the *spatial reuse gain* (see Figure 1.2). This simple example illustrates the concepts of collisions, energy-throughput trade-offs, and spatial reuse, which are present in every MAC scheme.

1.3.7   Modeling

The bases for analysis and simulations and analytical approaches are accurate and tractable models. Comprehensive network models should include the number of nodes and their relative distribution; their degree and type of mobility; the characteristics of the wireless link; the volume of traffic injected by the sources and the lifespan of their interaction; and detailed energy consumption models.

1.3.7.1   Wireless Link

An attenuation proportional to d^{α}, where d is the distance between two nodes and α is the so-called path loss exponent, is widely accepted as a model for path loss. Alpha ranges from 2 to 4 or even 5 [45], depending on the channel characteristics (environment, antenna position, frequency). This path loss model, together with the fact that packets are successfully transmitted if the signal-to-noise-and-interference ratio (SNIR) is bigger than some threshold [8], results in a deterministic model often used for analysis of multihop packet networks [23, 26, 41, 42, 46–48]. Thus, the radius for a successful transmission has a deterministic value, irrespective of the condition of the wireless channel. If only interferers within a certain distance of the receiver are considered, this "physical model" [23] turns into a "disk model".

The stochastic nature of the fading channel and thus the fact that the SINR is a random variable are mostly neglected. However, the volatility of the channel cannot be ignored in wireless networks [5, 8];

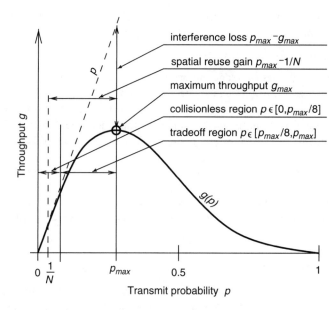

FIGURE 1.2 Generic throughput polynomial for a simple random MAC scheme.

Sousa and Silvester have also pointed out the inaccuracy of disk models [49] and it is easily demonstrated experimentally [50, 51]. In addition, this "prevalent all-or-nothing model" [52] leads to the assumption that a transmission over a multihop path fails completely or is 100% successful, ignoring the fact that end-to-end packet loss probabilities increase with the number of hops. Although fading has been considered in the context of packet networks [53, 54], its impact on the throughput of multihop networks and protocols at the MAC and higher layers is largely an open problem.

A more accurate channel model will have an impact on most of the metrics listed in Section 1.3.1. In the case of Rayleigh fading, first results show that the energy benefits of routing over many short hops may vanish completely, in particular if latency is taken into account [20, 55, 56]. The Rayleigh fading model not only is more accurate than the disk model, but also has the additional advantage of permitting separation of noise effects and interference effects due to the exponential distribution of the received power. As a consequence, the performance analysis can conveniently be split into the analysis of a zero-interference (noise-analysis) and a zero-noise (interference-analysis) network.

1.3.7.2 Energy Consumption

To model energy consumption, four basic different states of a node can be identified: transmission, reception, listening, and sleeping. They consist of the following tasks:

- *Acquisition:* sensing, A/D conversion, preprocessing, and perhaps storing
- *Transmission:* processing for address determination, packetization, encoding, framing, and maybe queuing; supply for the baseband and RF circuitry (The nonlinearity of the power amplifier must be taken into account because the power consumption is most likely not proportional to the transmit power [56].)
- *Reception:* Low-noise amplifier, downconverter oscillator, filtering, detection, decoding, error detection, and address check; reception even if a node is not the intended receiver
- *Listening:* Similar to reception except that the signal processing chain stops at the detection
- *Sleeping:* Power supply to stay alive

Reception and transmission comprise all the processing required for physical communication and networking protocols. For the physical layer, the energy consumption depends mostly on the circuitry, the error correction schemes, and the implementation of the receiver [57]. At the higher layers, the choice of protocols (e.g., routing, ARQ schemes, size of packet headers, number of beacons and other infra-structure packets) determines the energy efficiency.

1.3.7.3 Node Distribution and Mobility

Regular grids (square, triangle, hexagon) and uniformly random distributions are widely used analytically tractable models. The latter can be problematic because nodes can be arbitrarily close, leading to unrealistic received power levels if the path attenuation is assumed to be proportional to d^α. Regular grids overlaid with Gaussian variations in the positions may be more accurate. Generic mobility models for WSNs are difficult to define because they are highly application specific, so this issue must be studied on a case-by-case basis.

1.3.7.4 Traffic

Often, simulation work is based on constant bitrate traffic for convenience, but this is most probably not the typical traffic class. Models for bursty many-to-one traffic are needed, but they certainly depend strongly on the application.

1.3.8 Connectivity

Network connectivity is an important issue because it is crucial for most applications that the network is not partitioned into disjoint parts. If the nodes' positions are modeled as a Poisson point process in two dimensions (which, for all practical purposes, corresponds to a uniformly random distribution), the problem of connectivity has been studied using the tool of *continuum percolation theory* [58, 59]. For large networks, the phenomenon of a sharp phase transition can be observed: the probability that the network *percolates* jumps abruptly from almost 0 to almost 1 as soon as the density of the network is bigger than some critical value. Most such results are based on the geometric disk abstraction. It is conjectured, though, that other connectivity functions lead to better connectivity, i.e., the disk is apparently the hardest shape to connect [60]. A practical consequence of this conjecture is that fading results in improved connectivity. Recent work [61] also discusses the impact of interference. The simplifying assumptions necessary to achieve these results leave many open problems.

1.3.9 Quality of Service

Quality of service refers to the capability of a network to deliver data reliably and timely. A high *quantity of service*, i.e., throughput or transport capacity, is generally not sufficient to satisfy an application's delay requirements. Consequently, the *speed of propagation* of information may be as crucial as the throughput. Accordingly, in addition to network capacity, an important issue in many WSNs is that of quality-of-service (QoS) guarantees. Previous QoS-related work in wireless networks mostly focused on delay (see, for example, Lu et al. [62], Ju and Li [63], and Liu et al. [64]). QoS, in a broader sense, consists of the triple (R, P_e, D), where R denotes throughput; P_e denotes reliability as measured by, for example, bit error probability or packet loss probability; and D denotes delay. For a given R, the reliability of a connection as a function of the delay will follow the general curve shown in Figure 1.3

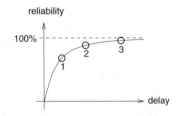

FIGURE 1.3 Reliability as a function of the delay. The circles indicate the QoS requirements of different possible traffic classes.

Note that capacity is only one point on the reliability-delay curve and therefore not always a relevant performance measure. For example, in certain sensing and control applications, the value of information quickly degrades as the latency increases. Because QoS is affected by design choices at the physical, medium-access, and network layers, an integrated approach to managing QoS is necessary.

1.3.10 Security

Depending on the application, security can be critical. The network should enable intrusion detection and tolerance as well as robust operation in the case of failure because, often, the sensor nodes are not protected against physical mishandling or attacks. Eavesdropping, jamming, and listen-and-retransmit attacks can hamper or prevent the operation; therefore, access control, message integrity, and confidentiality must be guaranteed.

1.3.11 Implementation

Companies such as Crossbow, Ember, Sensoria, and Millenial are building small sensor nodes with wireless capabilities. However, a per-node cost of $100 to $200 (not including sophisticated sensors) is prohibitive for large networks. Nodes must become an order of magnitude cheaper in order to render applications with a large number of nodes affordable. With the current pace of progress in VLSI and MEMS technology, this is bound to happen in the next few years. The fusion of MEMS and electronics onto a single chip, however, still poses difficulties. Miniaturization will make steady progress, except for two crucial components: the antenna and the battery, where it will be very challenging to find innovative solutions. Furthermore, the impact of the hardware on optimum protocol design is largely an open topic. The characteristics of the power amplifier, for example, greatly influence the energy efficiency of routing algorithms [56].

1.3.12 Other Issues

- *Distributed signal processing.* Most tasks require the combined effort of multiple network nodes, which requires protocols that provide coordination, efficient local exchange of information, and, possibly, hierarchical operation.
- *Synchronization and localization.* The notion of time is critical. Coordinated sensing and actuating in the physical world require a sense of global time that must be paired with relative or absolute knowledge of nodes' locations.
- *Wireless reprogramming.* A deployed WSN may need to be reprogrammed or updated. So far, no networking protocols are available to carry out such a task reliably in a multihop network. The main difficulty is the acknowledgment of packets in such a joint multihop/multicast communication.

1.4 Concluding Remarks

Wireless sensor networks have numerous exciting applications in virtually all fields of science and engineering, including health care, industry, military, security, environmental science, geology, agriculture, and social studies. In particular, the combination with macroscopic or MEMS-based actuators is intriguing because it permits manipulation of the environment in an unprecedented manner. Researchers and operators currently face a number of critical issues that need be resolved before these applications become reality. Wireless networking and distributed data processing of embedded sensing/actuating nodes under tight energy constraints demand new approaches to protocol design and hardware/software integration.

References

1. Sensor networks and applications, *IEEE Proc.*, 8, Aug. 2003.
2. Internet Engineering Task Force, Mobile ad-hoc networks (MANET). See http://www.ietf.org/html.charters/manet-charter.html.
3. Z.J. Haas et al., Eds., Wireless ad hoc networks, *IEEE J. Selected Areas Commun.*, 17, Aug. 1999. Special ed.
4. C.E. Perkins, Ed., *Ad Hoc Networking*. Addison Wesley, Reading, MA, 2000.
5. A.J. Goldsmith and S.B. Wicker, Design challenges for energy-constrained ad hoc wireless networks, *IEEE Wireless Commun.*, 9, 8–27, Aug. 2002.
6. A. Ephremides, Energy concerns in wireless networks, *IEEE Mag. Wireless Commun.*, 9, 48–59, Aug. 2002.
7. V. Rodoplu and T.H. Meng, Minimum energy mobile wireless networks, *IEEE J. Selected Areas Commun.*, 17(8), 1333–1344, 1999.
8. A. Ephremides, Energy concerns in wireless networks, *IEEE Wireless Commun.*, 9, 48–59, Aug. 2002.
9. Bluetooth wireless technology. Official Bluetooth site: http://www.bluetooth.com.
10. S. Tilak, N.B. Abu–Ghazaleh, and W. Heinzelman, A taxonomy of wireless micro-sensor network models, *ACM Mobile Computing Commun. Rev.*, 6(2), 28–36, 2002.
11. W.B. Heinzelman, A.P. Chandrakasan, and H. Balakrishnan, An application-specific protocol architecture for wireless microsensor networks, *IEEE Trans. Wireless Commun.*, 1, 660–670, Oct. 2002.
12. J. Kulik, W. Heinzelman, and H. Balakrishnan, Negotiation-based protocols for disseminating information in wireless sensor networks, *Wireless Networks*, 8, 169–185, March–May 2002.
13. A.B. McDonald and T.F. Znati, A mobility-based framework for adaptive clustering in wireless ad-hoc networks, *IEEE J. Selected Areas Commun.*, 17, 1466–1487, Aug. 1999.
14. S.S. Pradhan, J. Kusuma, and K. Ramchandran, Distributed compression in a dense microsensor network, *IEEE Signal Process. Mag.*, 19, 51–60, Mar. 2002.
15. A. Scaglione and S. Servetto, On the interdependence of routing and data compression in multi-hop sensor networks, in *Proc. ACM Int. Conf. Mobile Comp. Networks (MobiCom'02)*, Atlanta, GA, 140–147, Sept. 2002.
16. C. Intanagowiwat, R. Govindan, and D. Estrin, Directed diffusion: a scalable and robust communication paradigm for sensor networks, in *ACM Int. Conf. Mobile Computing Networking (MobiCom'00)*, Boston, MA, 56–67, Aug. 2000.
17. J.E. Wieselthier, G.D. Nguyen, and A. Ephremides, On the construction of energy-efficient broadcast and multicast trees in wireless networks, in *IEEE INFOCOM*, Tel Aviv, Israel, 585–594, Mar. 2000.
18. J.E. Wieselthier, G.D. Nguyen, and A. Ephremides, An insensitivity property of energy-limited wireless networks for session-based multicasting, in *IEEE ISIT*, Washington, D.C., June 2001.
19. J. Laneman, D. Tse, and G. Wornell, Cooperative diversity in wireless networks: efficient protocols and outage behavior, *IEEE Trans. Inf. Theory*. Accepted for publication. Available at: http://www.nd.edu/jnl/pubs/it2002.pdf.
20. M. Haenggi, A formalism for the analysis and design of time and path diversity schemes in wireless sensor networks, in *2nd Int. Workshop Inf. Process. Sensor Networks (IPSN'03)*, Palo Alto, CA, 417–431, Apr. 2003. Available at http://www.nd.edu/mhaenggi/ipsn03.pdf.
21. C.S. Raghavendra and S. Singh, PAMAS — power aware multi-access protocol with signaling for ad hoc networks, 1999. *ACM Computer Commun. Rev.* Available at: http://citeseer.nj.nec.com/460902.html.
22. C.-K. Toh, Maximum battery life routing to support ubiquitous mobile computing in wireless ad hoc networks, *IEEE Commun. Mag.*, 39, 138–147, June 2001.
23. P. Gupta and P.R. Kumar, The capacity of wireless networks, *IEEE Trans. Inf. Theory*, 46, 388–404, Mar. 2000.
24. P. Gupta and P.R. Kumar, Towards an information theory of large networks: an achievable rate region, in *IEEE Int. Symp. Inf. Theory*, Washington, D.C., 159, 2001.

25. L.-L. Xie and P.R. Kumar, A network information theory for wireless communication: scaling laws and optimal operation, Apr. 2002. submitted to *IEEE Trans. Inf. Theory*. Available at: http://black1.csl.uiuc.edu/prkumar/publications.html.

26. M. Grossglauser and D. Tse, Mobility increases the capacity of ad-hoc wireless networks, in *IEEE INFOCOM*, Anchorage, AL, 2001.

27. D. Tse and S. Hanly, Effective bandwidths in wireless networks with multiuser receivers, in *IEEE INFOCOM*, 35–42, 1998.

28. M. Gastpar and M. Vetterli, On the capacity of wireless networks: the relay case, in *IEEE INFOCOM*, New York, 2002.

29. G. Mergen and L. Tong, On the capacity of regular wireless networks with transceiver multipacket communication, in *IEEE Int. Symp. Inf. Theory*, Lausanne, Switzerland, 350, 2002.

30. S. Toumpis and A. Goldsmith, Capacity regions for wireless ad hoc networks, *IEEE Trans. Wireless Commun.*, 2, 736–748, July 2003.

31. D.N.C. Tse and S.V. Hanly, Multiaccess fading channels — part I: polymatroid structure, optimal resource allocation and throughput capacities, *IEEE Trans. Inf. Theory*, 44(7), 2796–2815, 1998.

32. S.V. Hanly and D.N.C. Tse, Multiaccess fading channels — part II: delay-limited capacities, *IEEE Trans. Inf. Theory*, 44(7), 2816–2831, 1998.

33. R. Negi and J.M. Cioffi, Delay-constrained capacity with causal feedback, *IEEE Trans. Inf. Theory*, 48, 2478–2494, Sept. 2002.

34. R.A. Berry and R.G. Gallager, Communication over fading channels with delay constraints, *IEEE Trans. Inf. Theory*, 48, 1135–1149, May 2002.

35. D. Tuninetti, On multiple-access block-fading channels, Mar. 2002. Ph.D. thesis, Institut EURE-COM. Available at: http://www.eurecom.fr/tuninett/publication.html.

36. J. Broch, D. Maltz, D. Johnson, Y. Hu, and J. Jetcheva, A performance comparison of multi-hop wireless ad hoc network routing protocols, in *ACM Int. Conf. Mobile Computing Networking (MobiCom)*, Dallas, TX, 85–97, Oct. 1998.

37. P. Johansson, T. Larsson, and N. Hedman, Scenario-based performance analysis of routing protocols for mobile ad-hoc networks, in *ACM MobiCom*, Seattle, WA, Aug. 1999.

38. S.R. Das, C.E. Perkins, and E.M. Royer, Performance comparison of two on-demand routing protocols for ad hoc networks, in *IEEE INFOCOM*, Mar. 2000.

39. C.R. Lin and J.-S. Liu, QoS Routing in ad hoc wireless networks, *IEEE J. Selected Areas Commun.*, 17, 1426–1438, Aug. 1999.

40. M. Haenggi, Energy-balancing strategies for wireless sensor networks, in *IEEE Int. Symp. Circuits Syst. (ISCAS'03)*, Bangkok, Thailand, May 2003. Available at http://www.nd.edu/mhaenggi/iscas03.pdf.

41. J.A. Silvester and L. Kleinrock, On the capacity of multihop slotted ALOHA networks with regular structure, *IEEE Trans. Commun.*, COM-31, 974–982, Aug. 1983.

42. L. Hu, Topology control for multihop packet networks, *IEEE Trans. Commun.*, 41(10), 1474–1481, 1993.

43. M. Haenggi, Probabilistic analysis of a simple MAC scheme for ad hoc wireless networks, in *IEEE CAS Workshop on Wireless Communications and Networking*, Pasadena, CA, Sept. 2002.

44. R.O. LaMaire, A. Krishna, and H. Ahmadi, Analysis of a wireless MAC protocol with client–server traffic and capture, *IEEE J. Selected Areas Commun.*, 12(8), 1299–1313, 1994.

45. T.S. Rappaport, *Wireless Communications — Principles and Practice*, 2nd ed., Prentice Hall, Englewood Cliffs, NJ.

46. H. Takagi and L. Kleinrock, Optimal transmission ranges for randomly distributed packet radio terminals, *IEEE Trans. Commun.*, COM-32, 246–257, Mar. 1984.

47. J.L. Wang and J.A. Silvester, Maximum number of independent paths and radio connectivity, *IEEE Trans. Commun.*, 41, 1482–1493, Oct. 1993.

48. C. Schurgers, V. Tsiatsis, S. Ganeriwal, and M. Srivastava, Optimizing sensor networks in the energy–latency–density design space, *IEEE Trans. Mobile Computing*, 1(1), 70–80, 2002.

49. E.S. Sousa and J.A. Silvester, Optimum transmission ranges in a direct-sequence spread-spectrum multihop packet radio network, *IEEE J. Selected Areas Commun.*, 8, 762–771, June 1990.

50. D.A. Maltz, J. Broch, and D.B. Johnson, Lessons from a full-scale multihop wireless ad hoc network testbed, *IEEE Personal Commun.*, 8, 8–15, Feb. 2001.

51. D. Ganesan, B. Krishnamachari, A. Woo, D. Culler, D. Estrin, and S. Wicker, An empirical study of epidemic algorithms in large scale multihop wireless networks, 2002. Intel Research Report IRB-TR-02-003. Available at www.intel-research.net/Publications/Berkeley/05022002170319.pdf.

52. T.J. Shepard, A channel access scheme for large dense packet radio networks, in *ACM SIGCOMM*, Stanford, CA, Aug. 1996. Available at: http://www.acm.org/sigcomm/sigcomm96/papers/shepard.ps.

53. M. Zorzi and S. Pupolin, Optimum transmission ranges in multihop packet radio networks in the presence of fading, *IEEE Trans. Commun.*, 43, 2201–2205, July 1995.

54. Y.Y. Kim and S. Li, Modeling multipath fading channel dynamics for packet data performance analysis, *Wireless Networks*, 6, 481–492, 2000.

55. M. Haenggi, On routing in random rayleigh fading networks, *IEEE Trans. Wireless Commun.*, 2003. Submitted for publication. Available at http://www.nd.edu/mhaenggi/routing.pdf.

56. M. Haenggi, The impact of power amplifier characteristics on routing in random wireless networks, in *IEEE Global Commun. Conf. (GLOBECOM'03)*, San Francisco, CA, Dec. 2003. Available at http://www.nd.edu/mhaenggi/globecom03.pdf.

57. H. Meyr, M. Moeneclaey, and S.A. Fechtel, *Digital Communication Receivers: Synchronization, Channel Estimation, and Signal Processing*. Wiley Interscience, 1998.

58. R. Meester and R. Roy, *Continuum Percolation*. Cambridge University Press, New York, 1996.

59. B. Bollobás, *Random Graphs*, 2nd ed. Cambridge University Press, New York, 2001.

60. L. Booth, J. Bruck, M. Cook, and M. Franceschetti, Ad hoc wireless networks with noisy links, in *IEEE Int. Symp. Inf. Theory*, Yokohama, Japan, 2003.

61. O. Dousse, F. Baccelli, and P. Thiran, Impact of interferences on connectivity in ad-hoc networks, in *IEEE INFOCOM*, San Francisco, CA, 2003.

62. S. Lu, V. Bharghavan, and R. Srikant, Fair scheduling in wireless packet networks, *IEEE/ACM Trans. Networking*, 7, 473–489, Aug. 1999.

63. J.-H. Ju and V.O.K. Li, TDMA scheduling design of multihop packet radio networks based on Latin squares, *IEEE J. Selected Areas Commun.*, 1345–1352, Aug. 1999.

64. H. Luo, S. Lu, and V. Bharghavan, A new model for packet scheduling in multihop wireless networks, in *ACM Int. Conf. Mobile Computing Networking (MobiCom'00)*, Boston, MA, 76–86, 2000.

65. H. Luo, P. Medvedev, J. Cheng, and S. Lu, A self-coordinating approach to distributed fair queueing in *ad hoc* wireless networks, *IEEE INFOCOM*, Anchorage, Apr. 2001.

66. A.E. Gamal, C. Nair, B. Prabhakar, E. Uysal-Biyikoglu, and S. Zahedi, Energy-efficient scheduling of packet transmissions over wireless networks, *IEEE INFOCOM*, New York, 2002, pp. 1773–1782.

67. S. Bhatnagar, B. Deb, and B. Nath, Service differentiation in sensor networks, *Fourth International Symposium on Wireless Personal Multimedia Communications*, Sept. 2001.

68. V. Kanodia, C. Li, A. Sabharwal, B. Sadeghi, and E. Knightly, Distributed multi-hop scheduling and medium access with delay and throughput constraints, *ACM MobiCom*, Rome, July 2001.

69. A. Striegel and G. Manimaran, Best-effort scheduling of (m, k)-firm real-time streams in multihop networks, *Workshop of Parallel and Distributed Real-Time Systems (WPDRTS) at IPDPS 2000*, Apr. 2000.

2

Next-Generation Technologies to Enable Sensor Networks*

Joel I. Goodman
MIT Lincoln Laboratory

Albert I. Reuther
MIT Lincoln Laboratory

David R. Martinez
MIT Lincoln Laboratory

2.1 Introduction

Several important technical advances make extracting more information from intelligence, surveillance, and reconnaissance (ISR) sensors very affordable and practical. As shown in Figure 2.1, for the radar application the most significant advancement is expected to come from employing collaborative and network centric sensor netting. One important application of this capability is to achieve ultrawideband multifrequency and multiaspect imaging by fusing the data from multiple sensors. In some cases, it is highly desirable to exploit multimodalities, in addition to multifrequency and multiaspect imaging.

Key enablers to fuse data from disparate sensors are the advent of high-speed fiber and wireless networks and the leveraging of distributed computing. ISR sensors need to perform enough on-board computation to match the available bandwidth; however, after some initial preprocessing, the data will be distributed across the network to be fused with other sensor data so as to maximize the information content. For example, on an experimental basis, MIT Lincoln Laboratory has demonstrated a virtual radar with ultrawideband frequency [1]. Two radars, located at the Lincoln Space Surveillance Complex

---

*This work is sponsored by the United States Air Force under Air Force contract F19628-00-C-002. Opinions, interpretations, conclusions, and recommendations are those of the authors and are not necessarily endorsed by the U.S. government.

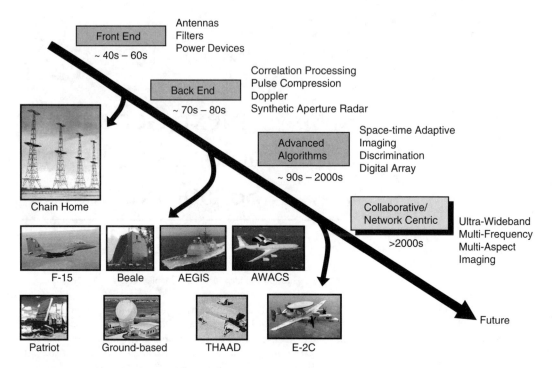

FIGURE 2.1 Radar technology evolution.

in Westford, Massachusetts, were employed; each of the two independent radars transmitted the data via a high-speed fiber network. The total bandwidth transmitted via fiber exceeded 1 Gbits/sec (billion bits per second). One radar was operating at X-band with 1-MHz bandwidth, and the second was operating at Ku-band with a 2-MHz bandwidth. A synthetic radar with an instantaneous bandwidth of 8 MHz was achieved after employing advanced ultrawideband signal processing [2].

These capabilities are now being extended to include high-speed wireless and fiber networking with distributed computing. As the Internet protocol (IP) technologies continue to advance in the commercial sector, the military can begin to leverage IP formatted sensor data to be compatible with commercial high-speed routers and switches. Sensor data from theater can be posted to high-speed networks, wireless and fiber, to request computing services as they become available on this network. The sensor data are processed in a distributed fashion across the network, thereby providing a larger pool of resources in real time to meet stringent latency requirements. The availability of distributed processing in a grid-computing architecture offers a high degree of robustness throughout the network. One important application to benefit from these advances is the ability to geolocate and identify mobile targets accurately from multiaspect sensor data.

2.1.1 Geolocation and Identification of Mobile Targets

Accurately geolocating and identifying mobile targets depends on the extraction of information from different sensor data. Typically, data from a single sensor are not sufficient to achieve a high probability of correct classification and still maintain a low probability of false alarm. This goal is challenging because mobile targets typically move at a wide range of speeds, tend to move and stop often, and can be easily mistaken for a civilian target. While the target is moving the sensor of choice is the ground moving target indication (GMTI). If the target stops, the same sensor or a different sensor working cooperatively must employ synthetic aperture radar (SAR). Before it can be declared foe, the target must often be confirmed with electro-optical or infrared (EO/IR) images. The goal of future networked systems is to have multiple sensors providing the necessary multimodality data to maximize the chances of accurately declaring a target.

A typical sensing sequence starts by a wide area surveillance platform, such as the Global Hawk unmanned aerial vehicle (UAV), covering several square kilometers until a target exceeds a detection threshold. The wide area surveillance will typically employ GMTI and SAR strip maps. Once a target has been detected, the on-board or off-board processing starts a track file to track the target carefully, using spot GMTI and spot SAR over a much smaller region than that initially covered when performing wide area surveillance. It is important to recognize that a sensor system is not merely tracking a single target; several target tracks can be going on in parallel. Therefore, future networked sensor architectures rely on sharing the information to maximize the available resources.

To date, the most advanced capability demonstrated is based on passing target detections among several sensors using the Navy cooperative engagement capability (CEC) system. Multisensor tracks are formed from the detection inputs arriving at a central location. Although this capability has provided a significant advancement, not all the information available from multimodality sensors has been exploited. The limitation is with the communication and available distributed computing. Multimodality sensor data together with multiple look angles can substantially improve the probability of correct classification vs. false alarm density. In addition to multiple modalities and multiple looks on the target, it is also desirable to send complex (amplitude and phase) radar GMTI data and SAR images to permit the use of high-definition vector imaging (HDVI) [3]. This technique permits much higher resolution on the target by suppressing noise around it, thereby enhancing the target image at the expense of using complex video data and much higher computational rates.

Another important tool to improve the probability of correct classification with minimal false alarm is high-range resolution (HRR) profiles. With this tool, the sensor bandwidth or, equivalently, the size of the resolution cell must be small resulting in a large data rate. However, it has been demonstrated that HRR can provide a significant improvement [4]. Therefore, next generation sensors depend on available communication pipes with enough bandwidth to share the individual sensor information effectively across the network. Once the data are posted on the network, the computational resources must exist to maintain low latencies from the time data become available to the time a target geoposition and identification are derived. The next subsection discusses the long-term architecture to implement netting of multiple sensor data efficiently.

2.1.2 Long-Term Architecture

In the future it will be desirable to minimize the infrastructure (foot print) forwardly deployed in the battlefield. It is most desirable to leverage high-speed satellite communication links to bring sensor data back to a combined air operations center (CAOC) established in the continental United States (CONUS).

The technology enablers for the long-term architecture shown in Figure 2.2 are high-speed, IP-based wireless and fiber communication networks, together with distributed grid computing. The in-theater commander's ability to task his organic resources to perform reconnaissance and surveillance of the opposing forces, and then to relay that information back to CONUS, allows significant reduction in the complexity, level, and cost of in-theater resources. Furthermore, this approach leverages the diverse analysis resources in CONUS, including highly trained personnel to support the rapid, accurate identification and localization of targets necessary to enable the time-critical engagement of surface mobile threats.

Space, air, and surface sensors will be deployed quickly to the battlefield. As shown in Figure 2.3, the stage in the processing chain at which the sensor data are tapped off to be sent via the network will dictate the amount of data transferred. For example, in a few applications one needs to send the data directly out of the analog-to-digital converters (A/D) to exploit coherent data combining from multiple sensors. Most commonly, it is preferable to perform on-board signal preprocessing to minimize the amount of data transferred. However, one must still be able to preserve content in the transferred data that is required to exploit features in the data not available from processing a signal sensor end to end. For example, one might be interested in transmitting wide area surveillance (WAS) data from SAR with high resolution to be followed by multiaspect SAR processing (shown in Figure 2.3 as application B). The data volume will be larger than the second example shown in Figure 2.3 as application A, in which

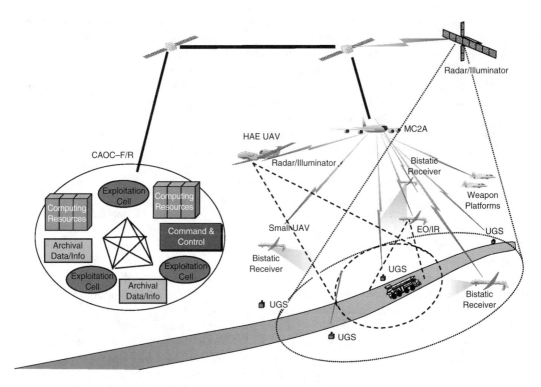

FIGURE 2.2 Postulated long-term architecture.

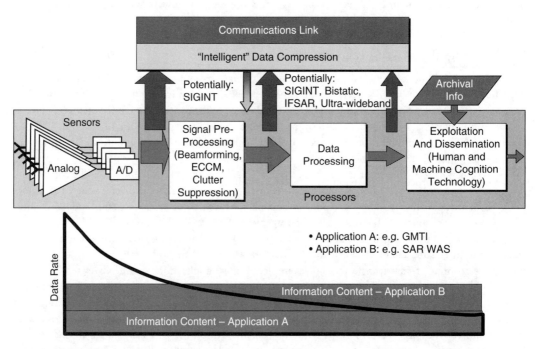

FIGURE 2.3 Sensor signal processing flow.

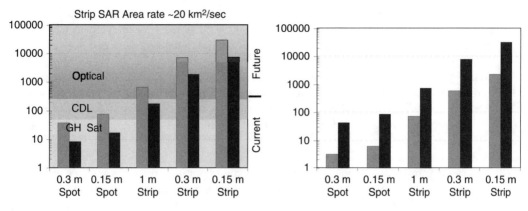

FIGURE 2.4 SAR data rate and computational throughput trade.

most of the GMTI processing is done on board. In any of these applications, it is paramount that "intelligent" data compression be done on board before data transmission to send only the necessary parts of the data requiring additional processing off board.

Each sensor will be capable of generating on-board processed data greater than 100 Mbits/sec (million bits per second). Figure 2.4 shows the trade-off between communication link data rates vs. on-board computation throughputs for different postulated levels of image resolution (for spot or strip map SAR modes). For example, for an assumed 1-m strip map SAR, one can send complex video radar data to then perform super-resolution processing off board. This approach would require sending between 100 to 1000 Mbits/sec. Another option is to perform the super-resolution processing on board, requiring between 100 billion floating-point operations per second (GFLOPS) to 1 trillion floating-point operations per second (TFLOPS).

Specialized military equipment, such as the common data link (CDL), can achieve data rates reaching 274 Mb/sec. If higher communication capacity were available, one would much prefer to send the large data volume for further processing off board to leverage information content available from multiple sensor data. As communication rates improve in the forthcoming years, it will not matter to the in-theater commander if the data are processed off board with the benefit of allowing exploitation of multiple sensor data at much rawer levels than is possible to date.

2.2 Goals for Real-Time Distributed Network Computing for Sensor Data Fusion

Several advantages can be gained by utilizing real-time distributed network computing to enable greater sensor data fusion processing. Distributed network computing potentially reduces the cost of the signal processing systems and the sensor platform because each individual sensor platform no longer needs as much processing capability as a stove-piped stand-alone system (although each platform may need higher bandwidth communications capabilities). Also, fault tolerance of the processing systems is increased because the processing and network systems are shared between sensors, thereby increasing the pool of available signal processors for all of the sensors. Furthermore, the granularity of managed resources is smaller; individual processors and network resources are managed as independent entities rather than managing an entire parallel computer and network as independent entities. This affords more flexible configuration and management of the resources.

To enable collaborative network processing of sensor signals, three technological areas are required to evolve and achieve maturity:

- Guaranteed *communication, storage buffer,* and *computation resources* must keep up with the high-throughput streams of data coming from the sensors. If any stage of the processing falls behind due to a network problem or interruption in the processor, buffering the data will become a problem quickly as increasing volumes of data must be stored to accommodate the delayed processors. Section 2.3 addresses technological possibilities to mitigate these resource availability issues.

- *Middleware* in the network of processors must be developed to accommodate a heterogeneous mix of computer and network resources. This middleware consists of a task control interface, which facilitates the communication between network resource management agents and entities, and an application programming interface for programming applications executed on the collaborative network processors. Section 2.4 will address these middleware interfaces.
- A *network resource manager* (NRM) system is necessary for orchestrating the execution of the application components on the computation and communication resources available in the collaborative network. Section 2.5 will discuss the components and functionality of the NRM.

2.3 The Convergence of Networking and Real-Time Computing

To date, networking of sensors has been demonstrated primarily using localized- and limited-capacity data links. As a result, the data available on the network from each sensor node typically represent the product of extensive prior processing of the radar data carried at the individual sensor. For example, the Navy CEC system, a relatively advanced current system, uses detection reports from independent sensors in the network to build composite tracks of targets. Access to raw (or possibly minimally preprocessed) multisensor data opens the opportunity for more effective exploitation of these data through integrated sensor data processing. The future network-centric ISR architecture will likely employ worldwide wideband communication networks to interconnect sensors with distributed processing and fusion sites. The resulting distributed database will provide a common operational picture for deployed forces. The sensor data will return to a CONUS entry point and pass over a wideband fiber network to the various processing centers where the sensor data will be fused. The data link from the theater to CONUS is expected to be optical to achieve very high link capacity [5].

This section discusses technologies that will guarantee that wireless and terrestrial network resources, storage buffer resources, and computational resources are available for sensor signal processing.

2.3.1 Guaranteeing Network Resources

Sensor data will traverse wireless and terrestrial (e.g., optical, twisted-copper) networks in which bit errors, packet loss, and delay could adversely affect the quality and timeliness of the ultimate result. The goal then is to choose a network and processing architecture to ameliorate the deleterious effects of data loss and network delay in the data fusion process. Due to the costs associated with developing, deploying, and maintaining a fixed terrestrial infrastructure, as well as inventing wholly new modulation protocols and standards for wireless and terrestrial signaling, it is cost-effective and expedient for military technology to ride the "commercial wave" of technical investment and progress in communication technologies.

With a fixed network infrastructure consisting primarily of commercial components, combating data loss and delay in terrestrial networks involves choosing the right protocols so that the network can enforce quality of service (QoS) demands; in wireless networks, this involves aggressive coding, modulation, and "lightweight" flow control for efficient bandwidth utilization. With sufficient complexity and bandwidth, it is possible with today's IP-based protocols to differentiate high-priority data to impart the mandated QoS for time-critical applications.

2.3.1.1 Terrestrial Networks

Reserving bandwidth on an IP-based network that is uniformly recognized across administrative domains involves employing protocols like RSVP-TE [6] or CR-LDP [7]. Although having sufficient communication bandwidth is an important aspect of processing sensor data in real time on a distributed network of resources, it does not guarantee real-time performance. For example, time-critical applications mapped onto networked resources should not have processing interrupted to service unmanaged traffic or be subject to a computational resource's resident operating system switching contexts to a lower priority task. For data that originate from sensors at very high streaming rates, a storage solution, as discussed in Section 2.3.2, is needed that is capable of recording sensor data in real time as well as robust in the face of network resource failures; this insures that a high-priority application can continue processing in the presence of malfunctioning or compromised networked equipment. However, adding a buffering storage solution only alleviates part of the problem; it does not mitigate the underlying problem of losing packets during network equipment failures or periods of network traffic that exceed network capacities.

For an IP-based network, one solution to this problem is to use remote agents deployed on primary compute resources or networked terminals located at switches that can dynamically filter unmanaged traffic. This is implemented by programming computer hardware specifically tasked with packet filtering (e.g., next generation gigabit Ethernet card) or dynamically reconfiguring the switch that directly connects to the compute resource in question by supplying an access control list (ACL) to block all packets except those associated with time-critical targeting. The formation of these exclusive networks using agents has been dubbed *dynamic private networks* (DPNs) — in effect, mechanisms for virtually overlaying a circuit switch onto a packet-switched network.

2.3.1.2　Wireless Networks

Unlike terrestrial networks, flow control and routing in mobile wireless sensor networks must contend with potentially long point-to-point propagation delays (e.g., satellite to ground) as well as a constantly changing topology. In a traditional terrestrial network employing link-state routing (e.g., OSPF), each node maintains a consistent view of a (primarily) fixed network topology so that a shortest path algorithm [8] can be used to find desirable routes from source to destination. This requires that nodes gather network connectivity information from other routers.

If OSPF were employed in a mobile wireless network, the overhead of exchanging network connectivity information about a transient topology could potentially consume the majority of the available bandwidth [9]. Routing protocols have been specifically designed to address the concerns of mobile networks [10]; these protocols fall into two general categories: proactive and reactive. Proactive routing protocols keep track of routes to all destinations, while reactive protocols acquire routes on demand. Unlike OSPF, proactive protocols do not need a consistent view of connectivity; that is, they trade optimal routes for feasible routes to reduce communication overhead. Reactive routes suffer a high initial overhead in establishing a route; however, the overall overhead of maintaining network connectivity is substantially reduced. The category of routing used is highly dependent upon how the sensors communicate with one another over the network.

Traditional flow control mechanisms over terrestrial networks that deliver reliable transport (e.g., TCP) may be inappropriate for wireless networks because, unlike wireless networks, terrestrial networks generally have a very low bit error rate (BER) on the order of 10^{-10}, so errors are primarily due to packet loss. Packet loss occurs in heavily congested networks when an ingress or egress queue of a switch or router begins to fill, requiring that some packets in the queue be discarded [11]. This condition is detected when acknowledgments from the destination node are not received by the source, prompting the source's flow control to throttle back the packet transmit rate [12].

In a wireless network in which BERs are four to five orders of magnitude higher than those of terrestrial networks, packet loss due to bit errors can be mistakenly associated with network congestion, and source flow control will mistakenly reduce the transmit rate of outgoing packets. Furthermore, when the source and destination are far apart, such as the communication between a satellite and ground terminal, where propagation delays can be on the order of 240 ms, delayed acknowledgments from the destination result in source flow control inefficiently using the available bandwidth. This is due to source flow control incrementally increasing the transmit rate as destination acknowledgements are received even though the entire frame of packets may have already been transmitted before the first packet reaches the receiver [13]. Therefore, to use bandwidth efficiently in a wireless network for reliable transport, flow control must be capable of differentiating BER from packet loss and account for long-haul packet transport by more efficiently using the available bandwidth. Some work in this area is reflected in RFC 2488 [14], as well as proposals for an explicit congestion warning, where, for example, the destination site would respond to packet errors with an acknowledgment that it received the source packets with a corruption notification.

At the physical layer, high data rates for a given BER have been realized by employing low-density parity check codes, such as turbo codes, in conjunction with bandwidth efficient modulation to achieve spectral efficiencies to within 0.7 dB of the Shannon limit [15]. Furthermore, extremely high spectral efficiencies have been demonstrated using multiple input, multiple output (MIMO) antenna systems whose theoretical channel capacity increases linearly with the number of transmit/receive antenna pairs [16]. Although turbo codes are advantageous as a forward error correction mechanism in wireless systems when trying to maximize throughput, MIMO systems achieve high spectral efficiencies only when operating in rich scattering environments [17]. In environments in which little scattering occurs, such as in some air-to-air communication links, MIMO systems offer very little improvement in spectral efficiency.

2.3.2   Guaranteeing Storage Buffer Resources

For a variety of reasons, it may be very desirable to record streaming sensor data directly to storage media while simultaneously sending the data on for immediate processing. For sensor signal processing applications, this enables multimodality data fusion of archived data with real-time (perishable) data from in-theatre sensors for improved target identification and visualization [18]. Storage media could also be used for rate conversion in cases in which the transmission rate exceeds the processing rate and for time-delay buffering for real-time robust fault tolerance (discussed in the next section). The storage media buffer reuse is deterministic and periodic so that management of the buffer is straightforward.

A number of possible solutions exist:

- *Directly attached storage* is a set of hard disks connected to a computer via SCSI or IDE/EIDE/ ATA; however, this technology does not scale well to the volume of streaming sensor data.
- *Storage area networks* are hard disk storage cabinets attached to a computer with a fast data link like Fibre Channel. The computer attached to the storage cabinet enjoys very fast access to data, but because the data must travel through that computer, which presents a single point of failure, to get to other computers on the network, this option is not a desirable solution.
- *Network-attached storage* connects the hard disk storage cabinet directly to the network as a file server. However, this technology offers only midrange performance, a single point of failure, and relatively high cost.

A visionary architecture in which data storage centers operate in parallel at a wide-area network (WAN) and local area network (LAN) level is described in Cooley et al. [19]. In this architecture, developed by MIT Lincoln Laboratory, high-rate streaming sensor data are stored in parallel across a partitioned network of storage arrays, which affords a highly scalable, low-cost solution that is relatively insensitive to communications or storage equipment failure. This system employs a novel and computationally efficient encoding and decoding algorithm using low-density parity check codes [20] for erasure recovery. Initial system performance measures indicate the erasure coding method described in Cooley et al. [19] has a significantly higher throughput and greater reliability when compared to Reed–Solomon, Tornado [21], and Luby [20] codes. This system offers a promising low-cost solution that scales in capability with the performance gains of commodity equipment.

2.3.3   Guaranteeing Computational Resources

The exponential growth in computing technology has contributed to making viable the implementation of advanced sensor processing in cost-effective hardware with form factors commensurate with the needs of military users. For example, several generations of embedded signal processors are shown in Figure 2.5.

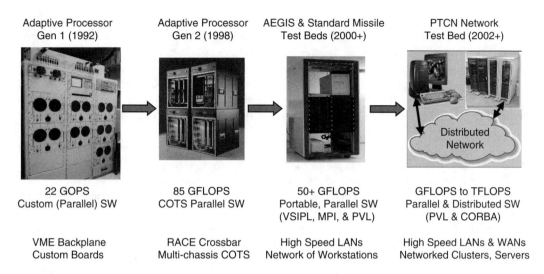

Adaptive Processor Gen 1 (1992)	Adaptive Processor Gen 2 (1998)	AEGIS & Standard Missile Test Beds (2000+)	PTCN Network Test Bed (2002+)
22 GOPS Custom (Parallel) SW	85 GFLOPS COTS Parallel SW	50+ GFLOPS Portable, Parallel SW (VSIPL, MPI, & PVL)	GFLOPS to TFLOPS Parallel & Distributed SW (PVL & CORBA)
VME Backplane Custom Boards	RACE Crossbar Multi-chassis COTS	High Speed LANs Network of Workstations	High Speed LANs & WANs Networked Clusters, Servers

FIGURE 2.5 Embedded signal processor evolution.

In the early 1990s, embedded signal processors were built using custom hardware and software. In the late 1990s, a move occurred from custom hardware to COTS processor systems running vendor-specific software together with application-specific parallel software tuned to each specific application. Most recently, the military embedded community is beginning to demonstrate requisite performance employing parallel and portable software running on COTS hardware.

Continuing technology advances in computation and communication will permit future signal processors to be built from commodity hardware distributed across a high-speed network and employing distributed, parallel, and portable software. These computing architectures will deliver 10^9 to 10^{12} floating point operations per second (GFLOPs to TFLOPs) in computational throughput. The distributed nature of the software will apply to on-board sensor processing as well as off-board processing. Clearly, on-board embedded processor systems will need to meet the stringent platform requirements in size, weight, and power.

Wireless and terrestrial network resources are not the only areas in which delays, failures, and errors must be avoided to process sensor data in a timely fashion. The system design must also guarantee that the marshaled compute nodes will keep up with the required computational throughput of streaming data at every stage of the processing chain. This guarantee encompasses two important facets: (1) keeping the processors from being interrupted while they are processing tasks and (2) implementing fail-over that is tolerant of fault.

2.3.3.1 Avoiding Processor Interruption

It is easy to take for granted that laptop and desktop computers will process commands as fast as the hardware and software are capable of doing so. A fact not generally known is that general computers are interrupted by system task processes and the processes of other applications (one's own and possibly from others working in the background on one's system). System task processes include keyboard and mouse input; communications on the Ethernet; system I/O; file system maintenance; log file entries; etc. When the computer interrupts an application to attend to such tasks, the execution of the application is temporarily suspended until the interrupting task has finished execution. However, because such interruptions often only consume a few milliseconds of processing time, they are virtually imperceptible to the user [22].

Nevertheless, the interruptions are detrimental to the execution of real-time applications. Any delay in processing these streams of data will instigate a need for buffering the data that will grow to insurmountable size as the delays escalate. A solution for these interrupt issues is to use a real-time operating system on the computation processors.

Simply put, real-time operating systems (RTOS) give priority to computational tasks. They usually do not offer as many operating system features (virtual memory, threaded processing, etc.) because of the interrupting processing nature of these features [22]. However, an RTOS can ensure that real-time critical tasks have guaranteed success in meeting streamed processing deadlines. An RTOS does not need to be run on typical embedded processors; it can also be deployed on Intel and AMD Pentium-class or Motorola G-series processor systems. This includes Beowulf clusters of standard desktop personal computers and commodity servers. This is an important benefit, providing a wide range of candidate heterogeneous computing resources.

A great deal of press has been generated in the past several years about real-time operating systems; however, the distinction between soft real-time and hard real-time operating systems is seldom discussed. Hard real-time systems guarantee the completion of tasks in a deterministic time period, while soft real-time systems give priority to critical tasks over other tasks but do not guarantee the completion of tasks in a deterministic time period [22]. Examples of hard real-time operating systems are VxWorks (Wind River Systems, Inc. [23]); RTLinux/Pro (FSMLabs, Inc. [24]); and pSOS (Wind River Systems, Inc. [23]), as well as dedicated massively parallel embedded operating systems like MC/OS (Mercury Computer Systems, Inc. [25]). Examples of soft real-time operating systems are Microsoft Pocket PC; Palm OS; certain real-time Linux releases [24, 26]; and others.

2.3.3.2 Working through System Faults

When fault tolerance in massively parallel computers is addressed, usually the solution is parallel redundant systems for fail-over. If a power supply or fan fails, another power supply or fan that is redundant in the system takes over the workload of the failed device. If a hard disk drive fails on a redundant array of independent disks (RAID) system, it can be hot swapped with a new drive and the contents of the drive rebuilt from the contents of the other drives along with checksum error correction code information. However, if an individual processor fails on a parallel computer, it is considered a failure of the entire parallel computer, and an identical backup computer is used as a fail-over. This backup system is then used as the primary computer, while the failed parallel computer is repaired to become the backup for the new primary eventually.

If, however, it were possible to isolate the failed processor and remap and rebind the processes on other processors in that computer — in real time — it would then be possible to have only a number of redundant processors in the system rather than entire redundant parallel computers. There are two strategies for determining the remapping as well as two strategies for handling the remapping and rebinding; each has its advantages and disadvantages.

To discuss these fail-over strategies, it is necessary to define the concepts of tasks and mappings. A signal processing application can be separated into a series of pipelined stages or tasks that are executed as part of the given application. A mapping is the task-parallel assignment of a task to a set of computer and network resources. In terms of determining the fail-over remapping, it is possible to choose a single remapping for each task or to choose a completely unique secondary path — a new mapping for each task that uses a set of processors mutually exclusive from the processors in the primary mapping path. If task backup mappings are chosen for each task, the fail-over will complete faster than a full processing chain fail-over; however, the rebinding fail-over for a failed task mapping is more difficult because the mappings from the task before and the task after the failed task mapping must be reconfigured to send data to and receive data from the new mapping. Conversely, if a completely unique secondary path is chosen as a fail-over, then fail-over completion will have a longer latency than performing a single task fail-over. However, the fail-over mechanics are simpler because the completely unique secondary path could be fully initialized and ready to receive the stream of data in the event of a failure in the primary mapping path.

In terms of handling the remapping and rebinding of tasks, it is possible to choose the fail-over mappings when the application is initially launched or immediately after a fault occurs. In either case, greater latency is incurred at launch time or after the occurrence of a fault. For these advanced options, support for this fault tolerance comes mainly from the middleware support, which is discussed in the next section, and from the NRM discussed in Section 2.5.

2.4 Middleware

Middleware not only provides a standard interface for communications between network resources and sensors for plug-and-play operation, but also enables the rapid implementation of high-performance embedded signal processing.

2.4.1 Control and Command of System

Because many systems use a diverse set of hardware, operating systems, programming languages, and communication protocols for processing sensor data, the manpower and time-to-deployment associated with integration have a significant cost. A middleware component providing a uniform interface that abstracts the lower-level system implementation details from the application interface is the common object request broker architecture (CORBA) [27]. CORBA is a specification and implementation that defines a standard interface between a client and server. CORBA leverages an interface definition language (IDL) that can be compiled and linked with an object's implementation and its clients. Thus, the CORBA standard enables client and server communications that are independent of the host hardware platforms, programming language, operating systems, and so on. CORBA has specifications and implementations to interface with popular communication protocols such as TCP/IP. However, this architecture has an open specification, general interORB protocol (GIOP) that enables developers to define and plug in platform-specific communication protocols for unique hardware and software interfaces that meet application-specific performance criteria.

For real-time and parallel embedded computing, it is necessary to interface with real-time operating systems, define end-to-end QoS parameters, and enact efficient data reorganization and queuing at communication interfaces. CORBA has recently included specifications for real-time performance and parallel processing, with the expectation that emerging implementations and specification addendums will produce efficient implementations. This will enable CORBA to move out of the command and control domain and be included as a middleware component involved in real-time and parallel processing of time-critical sensor data.

2.4.2 Parallel Processing

The ability to choose one of many potential parallel configurations enables numerous applications to share the same set of resources with various performance requirements. What is needed is a method to decouple the mapping, that is, the parallel instantiation of an application on target hardware, from generic serial application development. Automating the mapping process is the only feasible way of exploring the large parameter space of parallel configurations in a timely and cost-effective manner.

MIT Lincoln Laboratory has developed a C++-based library known as the parallel vector library (PVL) [28]. This library contains objects with parameterized methods deeply rooted in linear algebraic expressions commonly found in sensor signal processing. The parameters are used to direct the object instance to process data as one constituent part of a parallel whole. The parameters that organize objects in parallel configurations are run-time parameters so that new parallel configurations can be instantiated without having to recompile a suite of software. The technology of PVL is currently being incorporated into the parallel vector, signal, and image processing library for C++ (parallel VSIPL++) standard library [29].

2.5 Network Resource Management

Given the stated goals for distributed network computing for sensor fusion as outlined in Section 2.3, the associated network communication, storage, and processing challenges in Section 2.3, and the desire for standard interfaces and libraries to enable application parallelism and plug-and-play integration in Section 2.4, an integrated solution is needed that bridges network communications, distributed storage, distributed processing, and middleware. Clearly, it is possible for a development team to implement a

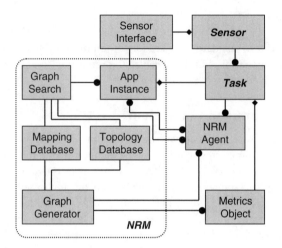

FIGURE 2.6 Object model for network resource manager (NRM).

"point" solution, but this is inherently not scalable and very difficult to maintain. Therefore an additional goal is to fully automate the process of configuring network communication, storage, and computational resources to process data for sensor fusion applications in real time, provide robust fault tolerance in the face of network resource failures, and impart this service in a highly dynamic network in the face of competing interests.

To address these needs, the network resource manager (NRM) was developed. The novelty and potency of the NRM is its capability of taking a sensor signal processing application designed and tested on single target processing element (PE) and mapping it in a task- and a data-parallel fashion across a network of computational resources to achieve real-time performance [30]. Figure 2.6 is an object-oriented model of the components that constitute the NRM. A high-level overview of the NRM follows, and details will be provided in the following subsections. The task of building a model from which the NRM launches parallel applications is broken into three distinct phases:

1. Map generation involves breaking an application into various task- and data-parallel components.
2. Map timing collects performance metric information associated with the components (or tasks) running on host resources. Using the performance metrics, the NRM creates a weighted graph-theoretic view of various permutations of an application mapped in parallel across networked resources.
3. Map selection finds the path through the graph that best meets system and application performance requirements.

The graph generator and graph search objects will heavily leverage PVL (discussed earlier) objects in the instantiation of task- and data-parallel configurations of applications on host resources. It should be noted, however, that the NRM's capabilities are fully general and independent from those of PVL and could work with other applications that are not developed using PVL to instantiate task- and data parallelism.

2.5.1 Graph Generator

As noted previously, PVL uses run-time parameters to generate new parallel configurations. This enables the NRM to launch applications in arbitrary parallel configurations using software developed for a single target PE without having to recompile the application software suite. The central challenge is to select a subset of the potentially astronomical number of permutations of parallel configurations as candidate parallel mappings. It is expected that the NRM will receive guidance in the form of performance and resource utilization bounds to help it avoid choosing undesirable configurations. It will also be given a

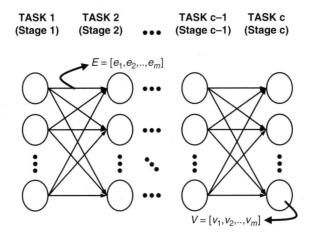

FIGURE 2.7 Sample graph with edge and vertex weights.

series of constituent tasks that comprise an application, so that its primary objective is to choose candidate data-parallel configurations for each of the individual tasks. Using a graph-theoretic model, the application space may be broken up as shown in Figure 2.7.

Each column in the graph is populated with vertices; each vertex corresponds to a mapping of the task corresponding to the given column to a potentially unique set of computational resources in the system. Each vertex has edges entering and exiting: entering edges correspond to communications with preceding tasks and exiting edges correspond to communications with succeeding tasks. Sensor signal processing applications may be represented as a stream signal processing flow, in which data move in one direction from task to task as they are processed. In this graph-theoretic model, task parallelism is represented along the horizontal axis of the graph, i.e., pipelined, overlapping execution intervals, while data parallelism is represented by the mapping of each task in the application onto one or more parallel computational resources of each vertex. The graph-theoretic representation of data- and task-parallel applications and the corresponding flow of communication enable the graph generator of the NRM to capture the potentially astronomical number of combinations of application-to-resource mappings in a concise and efficient fashion.

Finally, the graph generator is also responsible for launching the executable for each task mapping (vertex) on target resources so that performance metrics can be collected as discussed in the next subsection.

2.5.2 Metrics Object

The metrics object (MO) is responsible for collecting performance metrics of tasks launched by the graph generator. The MO works closely with the graph generator to weight the graph. Each of the resources that hosts a task is time synchronized; metric agents (see NRM agents in Subsection 2.5.4) on each of the resources will provide the MO measurements for it to formulate the following performance parameters associated with graph weights: throughput; latency; RAM memory; and PE utilization. The MO will calculate another metric known as processor cost, which is a ratio of compute horsepower used in the mapping to the overall processing horsepower available in the network.

Link utilization percentages within each mapping are also measured, as well as intertask utilization percentages. Map generation uses task column pairs to gather performance metrics in order to reduce the effort and time involved drastically. This is possible because the graph search algorithm will use a running tabulation of resource utilization percentages to ensure that simple linear superposition of path weights hold, given that these percentages remain under a given threshold. This is explained further in the next subsection. Once above the threshold, weight modifiers will be applied to subsequent stages during search. Finally, the metrics object will calculate a *network cost*, analogous to processor cost, which is a ratio of communications bandwidth used by a mapping pair with respect to the overall bandwidth available in the network.

2.5.3   Graph Search

The NRM must choose a path through the graph that determines the task mappings with which an application is launched on network resources. The choice of a path by the NRM is constrained by the time to result and the mandate to use a minimum set of networked resources. The data rate of the sensor data stream will drive required throughput for each task column in the graph; overall latency, which represents the total pipeline delay, is defined as the time period after which all data have been transmitted that a result is generated. To minimize any one application's impact on resource consumption, the path through the graph could be chosen to minimize the overall usage of computational or communication resources. This choice will depend upon whether an application is launched in a network that is compute resource or communication bandwidth limited.

The graph search problem may be formalized as a discrete and constrained optimization problem: given a set of hard constraints, minimize (or maximize) a given objective function. As described in the metrics object subsection, the NRM may choose constraints and an objective function from the set of weights shown in Table 2.1.

Scalar weights are singular — that is, only one is associated with a given vertex or edge; vector weights may include many elements in an edge or vertex association. Because each vertex and edge may represent the combination of many PE and network communication elements associated with a mapping pair, processor and network utilization may constitute weight vectors with many elements.

Although all weights tabulated previously may be chosen as constraints, memory, throughput, and network and PE utilization are not parameters that can be chosen as an objective function to optimize. This is because throughput is only a function of data rate; maximizing throughput has no impact on performance. Utilization also has no impact on performance and is only a measure of the validity of the solution. That is, subsequent stages in the graph may include resources from earlier stages, so keeping a running tabulation of utilization gives an indication of the onset of usage exceeding capacity and thereby degrading performance.

Network utilization and cost, PE utilization and cost, and memory are weights derived and constrained by the NRM, while data rate (throughput) and latency are application dependent and imposed by the sensor. The objective function that the NRM uses is chosen based on the desire to minimize an application's impact on resource usage or minimize the latency associated with an application's execution. For example, in a bandwidth-limited network, the graph search problem may be formulated as follows. While meeting application latency and throughput constraints, using less than 80% of the bandwidth available in the chosen network conduits and PEs and less than 100% of the available local PE-RAM memory, and using only a fraction of the overall processing bandwidth available network wide, select a parallel configuration for the

TABLE 2.1   Graph Weights
Associated with Individual Edges
and Vertices, and Corresponding
Sizes (Types)

Weight	Type
Latency	Scalar
Throughput	Scalar
PE utilization	Vector
Processor cost	Scalar
Network utilization	Vector
Network cost	Scalar
Memory	Scalar

application and the associated host resources using the smallest fraction of overall network bandwidth available. Even for moderately sized graphs (e.g., 1000 vertices by 10 stages), this is a complex combinatorial optimization problem; the general problem is NP complete. The authors have developed an iterative heuristic algorithm that has shown favorable performance for this class of problem in the quality of the solution and time to solution compared to other popular combinatorial optimization algorithms [31].

2.5.4 NRM Agents

The NRM agents are information and service links between the NRM and each of the resources. Agents must first register and be authenticated (e.g., using Kerberos [32]) before an NRM will invoke their services. This registration includes a characterization of the resource capabilities and services. When registered, the NRM will use these remotely deployed agents on computational resources to download and launch parameterized executables and modify the access control list (ACL) of switches and routers under its control in the formation of DPNs. Agents also provide a mechanism for centralized software maintenance and configuration by acting as transaction managers in the download and installation of applications, databases, middleware, etc. As stated earlier, the agents also provide a measurement object that is instantiated by applications to provide the NRM's MO with performance metrics during graph generation. Finally, agents give the NRM a view of the network state, periodically sending diagnostic messages indicating its operational status.

2.5.5 Sensor Interface

Sensors can be thought of as resources much like computational and communication resources, which are served by the NRM agents; thus, the sensor interface can be thought of as another type of NRM agent. Because many different sensor platforms could be served by an NRM-managed resource network, the sensor interface provides a common, abstract mechanism for communication between the NRM and the sensor platforms.

Sensors will request services through the sensor interface from the NRM using a well-defined middleware interface such as CORBA. This request for services involves requesting the proper application for the data stream that the sensor will be delivering to the network of resources as well as a request for the required metric constraints, such as throughput and latency (discussed in Subsection 2.5.2), needed to process the sensor data stream effectively. The determination of required constraints could involve negotiations between the sensor and the NRM through the sensor interface. The NRM uses the sensor interface to direct the sensor platform to start sending a data stream once the NRM has marshaled the resources that the sensor will need to satisfy the request. Finally, the sensor interface also facilitates communications between the sensor platform and the NRM regarding flow control, application shutdown, etc.

2.5.6 Mapping Database

This mapping database is populated with data structures generated by the graph generator and metrics object; it represents the weighted graph-theoretic characterization of the various parallel permutations of an application that is mapped to networked resources. Graph search uses the mapping database to reconstitute a weighted graph for each application for which it is asked to find resources and the degree and form of parallelism needed to meet real-time constraints.

2.5.7 Topology Database

The topology database stores the current state of each of the resources; the graph generator and graph search use this database. Graph generator uses the topology database to determine which resources are available and most appropriate for candidate task-application mappings. Graph search uses this database to verify that resources are functional before a set of resources is chosen to host an application, as well as for generating and modifying weights associated with resource utilization. The topology database is generated during the discovery phase when the NRM first comes online (e.g., see Breitbart et al. [33] and Astic and Foster [34]). Alternatively, an administrator could choose to generate a topology database for the NRM that enumerates connectivity and capability among all computation and storage resources under its control. Agent reports (or lack thereof) will affect state changes in this database indicating whether the resource is online or offline.

2.5.8   NRM Federation

In a large network with a sizeable number of resources, using a single NRM may not be the most effective solution. In such a scenario, multiple NRMs are organized in a bilevel hierarchy; wide-area network (WAN) NRMs interface with sensors and administer backbone communication resources, underneath which local-area network (LAN) NRMs administer and allocate compute resources for regional compute centers (RCCs). The primary responsibility of a WAN NRM is to choose a location on the network at which distributed computing is conducted for each application and to allocate WAN bandwidth for data flow between sensors and LAN resources. The objective of the WAN NRM is to load balance WAN traffic and computational load, taking into account the relative overall processing capability of each RCC. Each LAN NRM advertises its current processing capability using standardized metrics.

Each NRM is a federated collection, using a voting mechanism to elect an executor independently at the LAN and WAN levels. Each federation monitors the health of its executor by inspecting periodic diagnostic reports that the executor broadcasts. In response to an executor's diagnostic report (or lack thereof), the federation may choose to relieve the current executor of its responsibility and elect a new one. This prevents any one NRM failure from rendering resources unusable or disabling a sensor from contracting for network services.

Earlier paragraphs have detailed the LAN NRMs graph-theoretic representation of network resources, as well as its construction, weighting, and search criteria. The WAN NRM graph-theoretic representation and weighting are somewhat different from that of a LAN NRM; however, its construction and search criteria are formulated in an identical manner. The vertices in a WAN graph represent RCCs and each column corresponds to an application, while the concatenation of applications across the columns in a WAN NRM graph spans a mission. This is in contrast to a LAN NRM, in which the concatenation of tasks in its graph spans an application.

2.5.9   NRM Fault Tolerance

The absence of a heartbeat or the delivery of an error report by an agent alerts the NRM to a system fault. The NRM's fault tolerance policy is application dependent and is derived from a mandate by the developer and/or client. The policy is a trade-off between resource usage and seamless fail-over and includes redundant processing, surgical replacement, or restart of the application. Redundant processing is the most robust fail-over mechanism; the NRM simply assigns duplicate sets of resources to process the same data. If one set of resources fails, results are obtained from one of the duplicate sets. Redundant processing has the highest resource cost of all fault tolerant policies.

Conversely, the NRM may choose to replace the failed component dynamically so that processing is able to continue. In this case, the NRM may have allocated distributed network storage to act as a time-delay buffer in the event of resource failure. This would enable the application, if so instrumented, to pick up processing at the point at which the failure occurred. Finally, the NRM could simply choose to halt execution of the application and start over with a new set of processing resources, although a certain amount of data and the corresponding results may be lost irrevocably.

2.6   Experimental Results

A proof-of-concept experiment has been conducted at MIT Lincoln Laboratory in which the NRM allocates distributed networked resources for a sensor data fusion application in various scenarios [35].

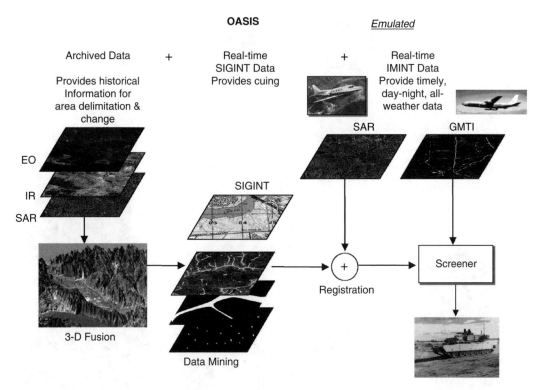

FIGURE 2.8 OASIS ATR and visualization.

TABLE 2.2 Synopsis of NRM Expected Performance

Experimental Configuration	Max Comm BW Requirement (MB/s)	Max Throughput Requirement (GFLOPS)	Processors Employed	Result Turn-Around Time
1 m data	26	0.7	1	1.6
1 m data with HDVI	26	2.2	2	2.6
1/4 m data	410	2.5	2	2.8
1/4 m data with HDVI	410	10	10	7

TABLE 2.3 Synopsis of NRM Performance

Experimental Configuration	Comm BW Measured (MB/s)	Throughput Measured (GFLOPS)	Processors Employed	Result Turn-Around Time
1 m data	26	0.7	1	1.4
1 m data with HDVI	26	2.2	2	2.5
1/4 m data	410	2.5	2	2.7
1/4 m data with HDVI	410	10	8	7.8

The sensor fusion application is OASIS (operator assisted integrated systems), which is an automatic target recognition and visualization suite (see Figure 2.8). OASIS processes real-time SAR data and archived data generated by sensors with different modalities like EO and IR [36]. A block diagram of the

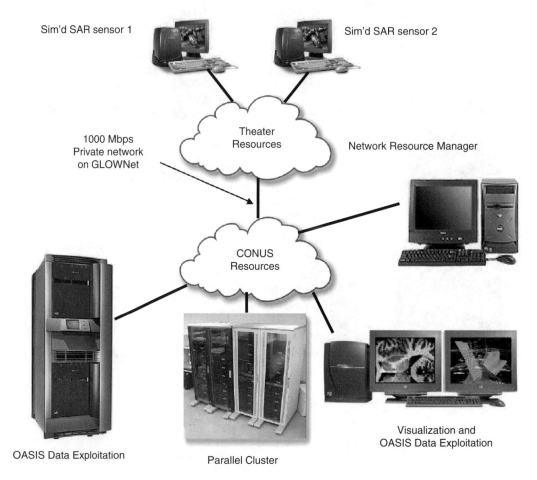

FIGURE 2.9 Experimentation resource network.

experimental test bed is shown in Figure 2.9. The experimentation resource network consisted of three SGI O2 workstations, an eight-processor SGI Origin, an eight-node, dual Pentium3 class Beowulf cluster, and a PC workstation, which hosted the NRM.

For this experiment, two SGI O2s were used as sensor surrogates to transmit unprocessed complex SAR imagery generated with range and cross-range resolutions of 1 and 1/4 m, respectively. The sensor surrogates fed data into the OASIS processing chain. To keep the complexity of the system manageable, only the most computationally intensive stage was made remappable. This stage, the HDVI processing [3] (stage 3 in Figure 2.10), had six options for the NRM ranging from a single SGI processor to six Pentium3 class cluster processors. The HDVI processing was conducted on targets detected on the two images at both resolutions, and image formation was conducted on processors in the local area network. The performance metrics for the OASIS applications were determined with a combination of actual performance measurements and modeled performance analyses. Table 2.2 is a tabulated synopsis of the expected performance of the NRM and Table 2.3 shows the actual performance of the NRM. The expected and actual performance values compared very well.

Because this network was PE resource limited, the objective of the NRM was to use the smallest fraction of PE bandwidth available across the network while meeting network conduit, PE utilization, latency, throughput, and network-wide bandwidth usage constraints. It is clear from the results that the NRM was able to tailor the communication and computation solution it delivered based on the particular

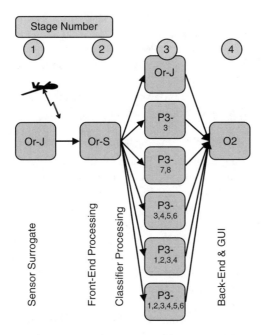

FIGURE 2.10 Graph of OASIS application onto the experimental resources.

application needs and the constraints imposed. The successful completion of this experiment has initiated further research and development to give the NRM greater functionality, automation, and flexibility.

Acknowledgments

The authors thank the members of the Precision Targeting via Collaborative Networking team at MIT Lincoln Laboratory for formulating many of the concepts discussed in this chapter. The authors also thank Dr. Mari Maeda, formerly of DARPA/ITO, and Dr. Gary Koob of DARPA/IPTO for their encouragement and support of this project.

References

1. Usoff, J., Beavers, W., and Cox, J., Wideband networked sensors processing, in *Proc. High Performance Embedded Computing Workshop*, November 2001.
2. Cuomo, K.M., Pion, J.E., and Mayhan, J.T., Ultrawide-band coherent processing, *IEEE Trans. Antenna Propagation*, 47, 1094, June 1999.
3. Benitz, G.R., High-definition vector imaging, *MIT Lincoln Lab. J., Special Issue Super-Resolution*, 10:2, 147, 1997.
4. Nguyen, D.H. et al., Super-resolution HRR ATR Performance with HDVI, *IEEE Trans. Aerospace Electron. Syst.*, 37:4, 1267, October 2001.
5. Chan, V.W.S., Optical space communications, *IEEE J. Selected Topics Quantum Electron.*, 6:6, 959, November/December, 2000.
6. Awduche, D. et al., RSVP-TE: extensions to RSVP for LSP tunnels, RFC 3209, http://www.faqs.org/rfcs/rfc3209.html, December 2001.
7. Ash, J. et al., Applicability statement for CR-LDP, RFC 3213, http://www.faqs.org/rfcs/rfc3213.html, January 2002.
8. Cormen, T.H., Leiserson, C.E., and Rivest, R.L., *Introduction to Algorithms*. McGraw–Hill, New York, 1993.

9. Strater, J. and Wollman, B., OSPF modeling and test results and recommendations, Mitre Technical Report 96W0000017, Mitre Corporation, 1996.

10. Perkins, C., *Ad Hoc Networking*, Addison–Wesley, Boston, 2001.

11. Floyd, S. and Jacobson, V., Random early detection gateways for congestion avoidance, *IEEE/ACM Trans. Networking*, 1:4, 397, August 1993.

12. Stevens, W., TCP slow start, congestion avoidance, fast retransmit and fast recovery algorithms, RFC 2001, http://www.faqs.org/rfcs/rfc2001.html, January 1997.

13. Stadler, J.S., Performance enhancements for TCP/IP on a satellite channel, in *Proc. IEEE Military Commun. Conf. 1998 (MILCOM98)*, 1, 270, October 1998.

14. Allman, M., Glover, D., and Sanchez, L., Enhancing TCP over satellite channels using standard mechanisms, RFC 2488, http://www.faqs.org/rfcs/rfc2488.html, January 1999.

15. Berrou, C., Glavieux, A., and Thitimajshima, P., Near Shannon limit error-correcting coding and decoding: turbo codes. 1, in *Conf. Rec. IEEE Int. Conf. Commun. 1993 (ICC 93)*, 2, 1064, May 1993.

16. Foschini, G.J., Layered space-time architecture for wireless communication in a fading environment when using multiple antennas, *Bell Labs Tech. J.*, 1:2, 41, Autumn 1996.

17. Raleigh, G.G. and Cioffi, J.M., Spatio-temporal coding for wireless communications, in *Proc. IEEE Global Telecommun. Conf. 1996 (GLOBECOM 96)*, 3, 1405, November 1996.

18. Sisterson, L.K. et al., An architecture for semi-automated radar image exploitation, *Lincoln Lab. J.*, 11:2, 175–204, 1998.

19. Cooley, J.A. et al., Software-based erasure codes for scalable distributed storage, in *Proc. 20th IEEE Symp. Mass Storage Syst.*, 157–164, April 2003.

20. Luby, M.G. et al., Practical loss-resilient codes, in *Proc. 29th ACM Symp. Theory Computing*, 150–159, 1997.

21. Byers, J.W., Luby, M.G., and Mitzenmacher, M., Accessing multiple mirror sites in parallel: using tornado codes to speed up downloads, in *Proc. IEEE INFOCOM 1999*, 275–283, March 1999.

22. Silberschatz, A. and Galvin, P., *Operating System Concepts*, 5th ed., Addison–Wesley, Reading, MA, 1998.

23. Wind River Systems, Inc. http://www.windriver.com/, accessed July 2003.

24. FSMLabs (Finite State Machine Labs), Inc. http://www.fsmlabs.com/, accessed July 2003.

25. Mercury Computer Systems, Inc. http://www.mc.com/, accessed July 2003.

26. Abbott, D., *Linux for Embedded and Real-Time Applications*, Newnes, Amsterdam, 2003.

27. Object Management Group. http://www.omg.org/, accessed July 2003.

28. Hoffmann, H., Kepner, J., and Bond, R., S3P: Automatic, optimized mapping of signal processing applications to parallel architectures, in *Proc. High Performance Embedded Computing Workshop 2001*, September 2001.

29. The vector, signal, and image processing library. http://www.vsipl.org/, accessed July 2002.

30. Reuther, A.I. and Goodman, J.I., Resource management for digital signal processing via distributed parallel computing, in *Proc. High Performance Embedded Computing Workshop 2002*, September 2002.

31. Goodman, J.I. et al., Discrete optimization using decision-directed learning for distributed networked computing, in *Proc. IEEE Asilomar Conf. Signal, Syst. Computers*, 1189–1196, November 2002.

32. Neuman, B.C. and Ts'o, T., Kerberos: an authentication service for computer networks, *IEEE Commun.*, 32:9, 33, September 1994.

33. Breitbart, Y. et al., Topology discover in heterogeneous IP networks, in *Proc. IEEE INFOCOM 2000*, 265–274, March 2000.

34. Astic, I. and Foster, O., A hierarchical topology discovery service for IPv6 networks, in *Proc. 2002 Network Operations Manage. Symp.*, 497–510, April 2002.

35. Reuther, A.I. and Goodman, J.I., dynamic resource management for a sensor-fusion application via distributed parallel grid computing, in *Proc. High Performance Embedded Computing Workshop 2003*, 2003.

36. Avent, R.K., A multi-sensor architecture for detecting high-value mobile targets, in *Proc. 2002 SIAM Conf. Imaging Sci. (IS02)*, March 2002.

3

A Taxonomy of Routing Techniques in Wireless Sensor Networks

Jamal N. Al-Karaki
Iowa State University

Ahmed E. Kamal
Iowa State University

3.1 Introduction

Wireless sensor networks (WSNs) contain hundreds or thousands of sensor nodes equipped with sensing, computing and communication abilities. Each node has the ability to sense elements of its environment, perform simple computations, and communicate among its peers or directly to an external base station (BS) (Figure 3.1). Deployment of a sensor network can be in random fashion (e.g., dropped from an airplane) or planted manually (e.g., fire alarm sensors in a facility). These networks promise a maintenance-free, fault-tolerant platform for gathering different kinds of data. Because a sensor node needs to operate for a long time on a tiny battery, innovative techniques to eliminate energy inefficiencies that would shorten the lifetime of the network must be used. A greater number of sensors allows for sensing over larger geographical regions with greater accuracy. The networking principles and protocols for WSNs are currently being investigated and developed [3–10]. Some application examples of WSNs include:

- Target field imaging
- Intrusion detection
- Weather monitoring
- Security and tactical surveillance
- Distributed computing
- Detecting ambient conditions such as temperature, movement, sound, light, or presence of certain objects
- Inventory control

Data sensing and reporting in sensor networks is dependent on the application and time criticality of the data reporting. As a result, sensor networks can be categorized as time-driven or event-driven

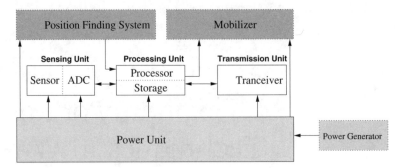

FIGURE 3.1 Components of a sensor node.

networks. The former type is suitable for applications that require periodic data monitoring. As such, sensor nodes will periodically switch on their sensors and transmitters, sense the environment, and transmit data of interest at constant periodic time intervals. Thus, they provide a snapshot of the relevant attributes at regular intervals. In the latter type, sensor nodes react immediately to sudden and drastic changes in the value of a sensed attribute due to the occurrence of a certain event. These are well suited for time critical applications.

A combination of these two types of communication is also possible. Moreover, WSNs can involve single-hop or multihop communication. In a single-hop WSN, a sensor node can directly communicate with any other sensor node or with the external base station. In multihop WSNs, however, communication between two sensor nodes may involve a sequence of hops through a chain of pairwise adjacent sensor nodes. A single-hop communication may take place between the base station and the sensor nodes, while the communication among the sensor nodes is typically multihop.

Despite the innumerable applications of WSNs, these networks have several restrictions, which should be considered when designing any protocol for these networks. Some of these limitations include:

- *Limited energy supply.* WSNs have a limited supply of energy; thus, energy-conserving communication protocols are necessary.
- *Limited computation.* Sensor nodes only have limited computing power, so WSNs cannot run a sophisticated network protocol.
- *Communication.* The bandwidth of the wireless links connecting sensor nodes is often limited, thus constraining the intersensor communication.

WSNs differ from traditional wireless networks like cellular networks in several ways. First, WSNs have severe energy constraints where the network needs to operate unattended for a long period of time. Second, in traditional wireless networks, the task of routing and mobility management is performed to optimize quality of service (QoS) and bandwidth efficiency; energy consumption is of secondary importance because the energy source can be replaced or recharged at any time. However, WSNs consist of nodes designed for unattended operation, so one task of routing is to optimize the use of energy so that the lifetime of the network is maximized. Third, nodes in WSNs are generally stationary after deployment except possibly for a few mobile nodes. Fourth, WSNs send redundant low-rate data in a many-to-one fashion.

MANETs and WSNs share some common problems. Among these are the time-varying characteristics of wireless links; limited power sources; possibility of link failures; scarce resources (e.g., bandwidth); multihop communications; and the ad hoc deployment of nodes in the network area. Although WSNs and MANETs involve multihop communications, the routing requirements are different in several ways:

- The destination in WSNs is known and communication is normally carried from multiple data sources to the BS (i.e., many to one); thus, the basic topology desired in data-gathering is a spanning tree. In MANETs, however, communication is generally on a peer–peer basis (i.e., one to one).

- Data collected by many sensors in WSNs are based on common phenomena, so there is a high probability that these data have some redundancy.
- MANETs are characterized by highly dynamic topologies due to free node mobility. In most application scenarios of WSNs, the sensors are not mobile and thus the nature of the dynamics is different.
- Mobile nodes in MANETs can have their energy sources (e.g., batteries) renewed, replaced, or recharged. The large number of sensor nodes, the necessity of unattended operation, and the long expected working lifetime of WSNs mean that the extremely limited energy resources must be managed carefully. Moreover, limited energy resources, in turn, preclude high data rate communication in WSNs.

The aforementioned reasons make the many end-to-end routing schemes proposed for MANETs in the literature inappropriate for WSNs under these conditions.

3.1.1 Motivation and Design Issues in WSN Routing

One of the main design goals of WSNs is to prolong the lifetime of the network and prevent connectivity degradation by employing aggressive energy management techniques. This is motivated by the fact that energy sources in WSNs are irreplaceable and their lifetime is limited. However, the positions of the sensor nodes are usually not engineered or predetermined and thus allow random deployment in inaccessible terrain or disaster relief operations. This implies that the nodes are expected to perform sensing and communication with no continual maintenance or human attendance and battery replenishment, which limits the amount of energy available to the sensor nodes. Therefore, extensive collaboration between sensor nodes is required to perform high-quality sensing and to behave as fault-tolerant systems. Current routing protocols designed for traditional networks cannot be used directly in a sensor network because:

- Sensor nodes should be self-organizing because the ad hoc deployment of these nodes requires the system to form connections and cope with the resultant distribution. The operation of the sensor networks is unattended, so network organization and configuration should be performed automatically.
- In most application scenarios, sensor nodes are stationary. However, in some applications, some sensor nodes may be allowed to move and change their location (though very low mobility).
- Sensor networks are application specific (i.e., design requirements of a sensor network change with application). For example, the challenging problem of low-latency precision tactical surveillance is different from that required for a periodic weather-monitoring task.
- Data collected by many sensors in WSNs are based on common phenomena; there is a high probability that these data have some redundancy (i.e., data redundancy). Therefore, in-network aggregation of data is needed to yield energy-efficient data delivery before dispatch to destinations. Data redundancy may consume sensor nodes' energy as a result of unnecessary and replicated transmissions.
- Sensor networks are data-centric networks. In traditional networks, data are requested from a specific node. In sensor networks, data are requested based on certain attributes. The sensors can remain in the sleep state, with the data reported from the few remaining sensors providing lower quality. Once an event of interest is detected, the system should be able to configure so as to obtain very high-quality results.
- WSNs have relatively large numbers of sensor nodes, potentially on the order of thousands of nodes. Therefore, sensor nodes need not have a unique ID because the overhead of ID maintenance is high. In data-centric WSNs, the data can be more important than knowing which nodes sent the data.
- WSNs use attribute-based addressing. A user issues an attribute-based address composed of a set of attribute–value pair query. For example, if the query is [temperature > 60°F], then sensor nodes that sense temperature > 60°F only need to respond and report their readings.

- Position awareness of sensor nodes is important because data collection is based on the location. Currently, it is not feasible to use global positioning system (GPS) hardware for this purpose. Methods based on triangulation [14], for example, allow sensor nodes to approximate their position using radio strength from a few known points. Bulusu and colleagues [14] have found that algorithms based on triangulation can work quite well under conditions in which only a very few nodes know their positions *a priori*, e.g., using GPS hardware. Nevertheless, it is favorable to have GPS-free solutions [15] for the location problem in WSNs.

Effective design and deployment of efficient routing protocols in WSNs still face several challenges. These are discussed briefly in the next section.

3.1.2   Routing Challenges in WSNs

The design of routing protocols in WSNs is influenced by many challenging factors that must be overcome before efficient communication can be achieved in WSNs. Some of these challenges and some design guidelines to be considered in the design process include:

- *Ad hoc deployment*. Sensor nodes are deployed randomly. This requires that the system be able to cope with the resultant distribution and form connections between the nodes. Thus, the system should be adaptive to changes in network connectivity as a result of node failure.
- *Energy consumption without losing accuracy*. Sensor nodes can use up their limited supply of energy performing computations and transmitting information in a wireless environment. As such, energy-conserving forms of communication and computation are essential. Sensor node lifetime shows a strong dependence on battery lifetime. In a multihop WSN, each node plays a dual role as data sender and data router. The malfunctioning of some sensor nodes because of power failure can cause significant topological changes and might require rerouting packets and reorganizing the network.
- *Computation capabilities*. Sensor nodes have limited computing power and therefore may not be able to run sophisticated network protocols. Therefore, new or light-weight and simple versions of traditional routing protocols are needed to fit in the WSN environment.
- *Communication range*. Intersensor communication exhibits short transmission ranges. Therefore, it is most likely that a route will generally consist of multiple wireless hops.
- *Fault tolerance*. Some sensor nodes may fail or be blocked due to lack of power, physical damage, or environmental interference. The failure of sensor nodes should not affect the overall task of the sensor network. If many nodes fail, MAC and routing protocols must accommodate formation of new links and routes to the data collection base stations. This may require actively adjusting transmit powers and signaling rates on the existing links to reduce energy consumption, or rerouting packets through regions of the network where more energy is available. Therefore, multiple levels of redundancy may be needed in a fault-tolerant sensor network.
- *Scalability*. The number of sensor nodes deployed in the sensing area may be in the order of hundreds or thousands or more. Any scheme must be able to work with this huge number of sensor nodes. Also, change in network size, node density, and topology should not affect the task and operation of the sensor network. In addition, sensor network routing protocols should be scalable enough to respond to events in the environment. Until an event occurs, most of the sensors can remain in the sleep state, with data from the few remaining sensors providing a coarse quality. Once an event of interest is detected, the system should be able to configure so as to obtain very high-quality results.
- *Hardware constraints*. Consisting of many hardware components, a sensor node may be smaller than a cubic centimeter. These components consume extremely low power and operate in an unattended mode; nonetheless, they should adapt to the environment of the sensor network and function correctly.

- *Transmission media.* In a multihop sensor network, communicating nodes are linked by a wireless medium. The traditional problems associated with a wireless channel (e.g., fading, high error rate) may also affect the operation of the sensor network. In general, the required bandwidth of sensor data will be low, on the order of 1 to 100 kb/s. Related to the transmission media is the design of medium access control (MAC). One approach of MAC design for sensor networks is to use TDMA-based protocols that conserve more energy compared to contention-based protocols like CSMA (e.g., IEEE 802.11). However, although TDMA-based protocols work fine in a flat network, they do not adapt well to clustered WSNs. Management of intercluster communication and dynamic adaptation of the TDMA protocol to variation in the number of nodes in the cluster — in terms of its frame length and time slot assignment — are key challenges for the MAC protocol in hierarchical network. In WSNs, sensors use the Bluetooth technology for transmission. Bluetooth is based upon low-cost, low-complexity, and short range radio communication of data and voice in stationary and mobile environments.
- *Connectivity.* High node density in sensor networks precludes their complete isolation from each other. Therefore, sensor nodes are expected to be highly connected. This, however, may not prevent the network topology from being variable and the network size from being changed due to sensor nodes' failures for different reasons.
- *Control overhead.* When the number of retransmissions in a wireless medium increases due to collisions, latency and energy consumption will also increase. Therefore, control packet overhead increases linearly with node density. As a result, trade-offs among energy conservation, self-configuration, per-node fairness, and latency may exist. However, fairness and throughput are of secondary importance in WSNs.
- *Quality of service.* In some applications, the data should be delivered within a certain period of time from the moment they are sensed; otherwise the data will be useless. Therefore, bounded latency for data delivery is another condition for time-constrained applications.

The communication architecture of the sensor network is shown in Figure 3.2. The sensor nodes are usually scattered in a sensor field — an area in which the sensor nodes are deployed. The nodes in these networks coordinate to produce high-quality information about the physical environment. Each sensor

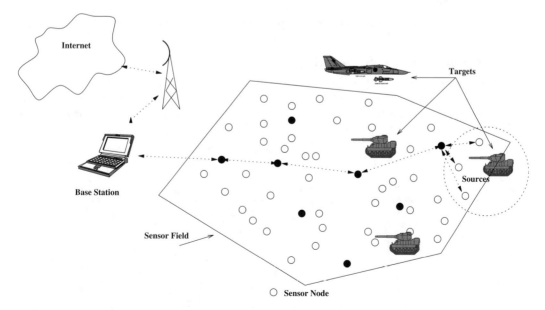

FIGURE 3.2 Communication architecture of a sensor network.

node bases its decisions on its mission, the information it currently has, and its knowledge of its computing, communication and energy resources. Each of these scattered sensor nodes has the capabilities to collect data and route data back to the base stations. A base station may be a fixed node or a mobile node capable of connecting the sensor network to an existing communications infrastructure or to the Internet where a user can have access to the reported data.

3.2 Routing Protocols in WSNs

In sensor networks, conservation of energy, which is directly related to network lifetime, is considered relatively more important than the performance of the network in terms of quality of data sent. As the energy gets depleted, the network may be required to reduce the quality of the results in order to reduce the energy dissipation in the nodes and thus lengthen total network lifetime. Therefore, conservation of energy is considered to be more important than the performance of the network.

Recently, routing protocols for WSNs have been extensively studied. In general, routing in WSNs can be divided into flat-based routing, hierarchical-based routing, and adaptive-based routing. In flat-based routing, all nodes are assigned equal roles. In hierarchical-based routing, however, nodes will play different roles in the network. In adaptive routing, certain system parameters are controlled in order to adapt to the network's current conditions and available energy levels. Furthermore, these protocols can be classified into multipath-based, query-based, or negotiation-based routing techniques depending on the protocol operation. In order to streamline this survey, classification according to the network structure and routing criteria is used. The classification is shown in Figure 3.3. Note that because the topology is static, it is preferable to have a table-driven routing protocol because a lot of energy is used in route discovery and setup of reactive protocols. Another class of routing protocols is the cooperative routing protocols in which nodes send the data to a central node at which data can be aggregated and may be subject to further processing. Therefore, reducing route cost in terms of energy use is of great importance.

Several energy-aware routing protocols have been proposed to capture this requirement. The rest of this section presents a detailed overview of the main routing paradigms in WSNs.

3.2.1 Flat Routing

The first category of routing protocols is the multihop flat routing protocols, summarized in the remainder of this subsection.

3.2.1.1 Sequential Assignment Routing (SAR)

Routing decision in SAR [11] is dependent on three factors: energy resources, QoS on each path, and the priority level of each packet. To avoid single-route failure, a multipath approach and localized path restoration schemes are used. To create multiple paths from a source node, a tree rooted at the source

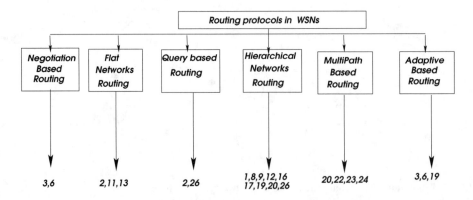

FIGURE 3.3 Routing protocols in WSNs: a taxonomy.

node to the destination nodes (i.e., the set of base stations) is built. The paths of the tree are built while avoiding nodes with low energy or QoS guarantees. At the end of this process, each sensor node will be part of the multipath tree.

For each node, two metrics are associated with each path: an additive QoS metric, i.e., delay, and a measure of the energy usage for routing on that path. The energy is measured with respect to how many packets will traverse that path. SAR will calculate a weighted QoS metric as the product of the additive QoS metric and a weight coefficient associated with the priority level of the packet. The objective of the SAR algorithm is to minimize the average weighted QoS metric throughout the lifetime of the network. If topology changes due to node failures, a path recomputation is needed. As a preventive measure, a periodic recomputation of paths is triggered by the base station to account for any changes in the topology. A handshake procedure based on a local path restoration scheme between neighboring nodes is used to recover from a failure.

3.2.1.2 Directed Diffusion

Intanagonwiwat et al. [2] have presented a data-centric and application-aware paradigm called directed diffusion. It is data centric (DC) in the sense that all the data generated by sensor nodes are named by attribute–value pairs. DC performs in-network aggregation of data to yield energy-efficient data delivery. The main idea of the DC paradigm is to combine the data coming from different sources en route — eliminating redundancy, minimizing the number of transmissions, and thus saving network energy and prolonging its lifetime.

This paradigm is different from the traditional paradigm, termed address centric (AC). In AC routing, the problem is to find short routes between pairs of addressable mobile nodes (end-to-end routing); DC finds routes from multiple sources to a single destination that allow in-network consolidation of redundant data. Figure 3.4 shows an example of the difference between address-centric and data-centric routing. In Figure 3.4(a) is an example of AC routing in which three source nodes detect a target and each uses an end-to-end path independently of the others to report data to the sink node. Using DC routing (Figure 3.4b), an aggregated form of the data received by node B is sent to the sink node, resulting in less energy expenditure.

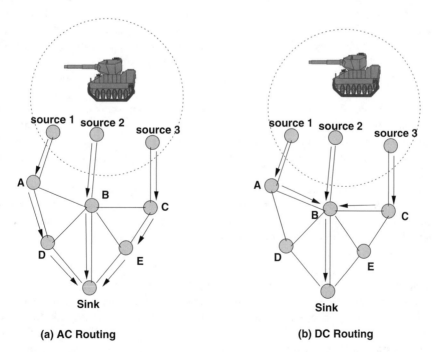

(a) AC Routing   (b) DC Routing

FIGURE 3.4 Differences between (a) address-centric (AC) and (b) data-centric (DC) routing.

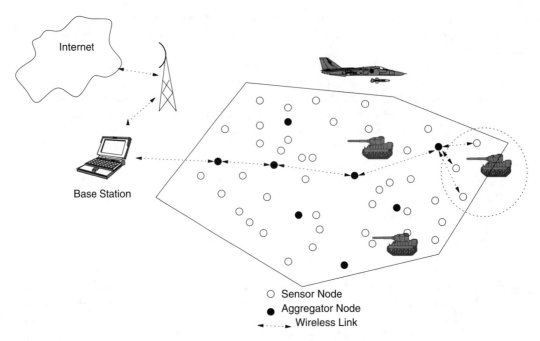

FIGURE 3.5 Sensor network used in military application and employing directed diffusion. A set of sensor nodes (black circles) are selected to work as data aggregators; through them data are sent to the external base station. If an Internet connection is available, a quality copy of the readings can be sent through the Internet to the central command, for example.

The application of this paradigm to query dissemination and processing has been demonstrated in Intanagonwiwat et al. [2]. The query is disseminated or flooded throughout the network and gradients are set up to draw data satisfying the query toward the requesting node; that is, a sink may query for data by disseminating interests and intermediate nodes propagate these interests. More generally, a gradient specifies an attribute value and a direction. Events (i.e., data) start flowing toward the requesting node from multiple paths. A small number of paths can be reinforced so as to prevent further flooding according to a local rule. Then an empirically low delay path is selected to be reinforced. The strength of the gradient may be different toward different neighbors, resulting in different amounts of information flow (see Figure 3.5, for example).

Another use of directed diffusion is to propagate an important event spontaneously to some sections of the sensor network. This type of information retrieval is well suited only for persistent queries in which requesting nodes are not expecting data that satisfy a query for duration of time. This makes it unsuitable for one-time queries because it is not worth setting up gradients, etc. for queries that employ the path only once.

Interest describes a task required to be done by the sensor net. Interest is injected at some point, normally at BS; the source is unknown at this point. Interest diffuses through the network hop by hop and is broadcast by each node to its neighbors. At this stage, loops are not checked for; they are removed at a later stage. Figure 3.6 shows an example of the working of directed diffusion (sending interests, building gradients, and data dissemination).

All sensor nodes in a directed diffusion-based network are application-aware, which enables diffusion to achieve energy savings by selecting empirically good paths and by caching and processing data in the network. In a sensor network based on directed diffusion, each sensor node names data that it generates with one or more attributes. The sink broadcasts the interest, which is a named task descriptor, to all sensors. The task descriptors are named by assigning attribute–value pairs that describe the task. Each sensor node then stores the interest entry in its cache. The interest entry contains a time stamp field and

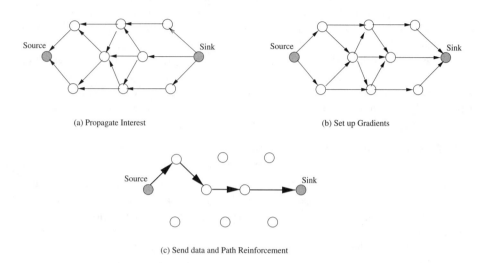

(a) Propagate Interest   (b) Set up Gradients

(c) Send data and Path Reinforcement

FIGURE 3.6 Interest diffusion in a sensor network.

several gradient fields. As the interest is propagated throughout the network, the gradients from the source back to the sink are set up.

Caching can increase the efficiency, robustness, and scalability of coordination between sensor nodes, which is the essence of the data diffusion paradigm. Locally cached data may be accessed by other users with lower energy consumption than if the data were to be resent end to end. When the source has data for the interest, the source sends the data along the interest's gradient path. As the data propagates, data may be transformed locally at each node. The sink periodically refreshes and resends the interest when it starts to receive data from the source. This is necessary because interests are not reliably transmitted throughout the network. The main goal of this protocol is to compute a path robustly from source to sink through the use of attribute-based naming and gradient paths.

The performance of data aggregation methods used in the directed diffusion paradigm is affected by the positions of the source nodes in the network, the number of sources, and the communication network topology. In order to investigate these factors, two models of source placement, called the event radius (ER) model and the random source (RS) model (shown in Figure 3.7), were studied. In the ER model,

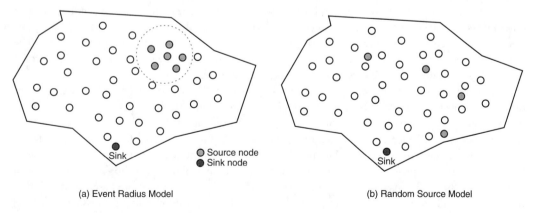

(a) Event Radius Model   (b) Random Source Model

FIGURE 3.7 Two models used in data-centric routing.

a single point in the network area is defined as the location of an event. This may correspond to a vehicle or some other phenomenon tracked by the sensor nodes. All nodes within a distance S (called the sensing range) of this event that are not sinks are considered to be data sources. The average number of sources is approximately $\pi S^2 n$ for a network with n nodes.

In the RS model, k of the nodes that are not sinks are randomly selected to be sources. Unlike the ER model, in RS the sources are not necessarily clustered near each other. In both models of source placement, for a given energy budget, a greater number of sources can be connected to the sink. Thus, the energy savings with aggregation used in the directed diffusion can be transformed to provide a greater degree of robustness to dynamics in the sensed phenomena.

3.2.1.3 Minimum Cost Forwarding Algorithm

The minimum cost forwarding algorithm (MCFA) [13] exploits the fact that the direction of routing is always known (i.e., toward the fixed external base station). Thus, a sensor node need not have a unique ID or maintain a routing table. Instead, each node maintains the least cost estimate from itself to the base station. Each message to be forwarded by the sensor node is broadcast to its neighbors. When a node receives the message, it checks if it is on the least cost path between the source sensor node and the base station. If this is the case, it rebroadcasts the message to its neighbors. This process repeats until the base station is reached. In MCFA, each node should know the least cost path estimate from itself to the base station. This is obtained as follows.

The base station broadcasts a message with the cost set to zero while every node initially sets its least cost to the base station to infinity (∞). Each node, upon receiving the broadcast message originated at the base station, checks to see if the estimate in the message plus the link on which it is received are less than the current estimate. If so, the current estimate and the estimate in the broadcast message are updated. If the received broadcast message is updated, then it is resent; otherwise, it is purged and nothing further is done. However, the previous procedure may result in some nodes having multiple updates and nodes far away from the base station will get more updates from those closer to the base station. To avoid this, the MCFA was modified to run a backoff algorithm at the setup phase. The backoff algorithm dictates that a node will not send the updated message until $a*l_c$ time units have elapsed from the time at which the message is updated, where a is a constant and l_c is the link cost from which the message was received.

3.2.1.4 Coherent and Noncoherent Processing

Data processing is a major component in the operation of wireless sensor networks. Thus, routing techniques employ different data processing techniques. In general, sensor nodes will cooperate with each other in processing different data flooded in the network area. Two examples of data processing techniques proposed in WSNs are coherent and noncoherent data processing-based routing [11]. In noncoherent data processing routing, nodes will locally process the raw data before sending them to other nodes for further processing. The nodes that perform the further processing are called the aggregators. In coherent routing, the data are forwarded to aggregators after minimum processing. The minimum processing typically includes tasks like time stamping, duplicate suppression, etc. To perform energy-efficient routing, coherent processing is normally selected.

Noncoherent functions have fairly low data traffic loading. On the other hand, because coherent processing generates long data streams, energy efficiency must be achieved by path optimality. Noncoherent cooperative processing contains three phases in the processing: (1) target detection, data collection, and preprocessing; (2) membership declaration; and (3) central node election. During phase 1, a target is detected and its data are collected and preprocessed. When a node decides to participate in a cooperative function, it will enter phase 2 and declare this intention to all neighbors. This should be done as soon as possible so that each sensor has a local understanding of the network topology. Phase 3 is the election of the central node, which is selected to perform more sophisticated information processing; therefore, it must have sufficient energy reserves and computational capability.

Sohrabi and Pottie [11] proposed single and multiple winner algorithms for noncoherent and coherent processing, respectively. In the single winner algorithm (SWE), a single aggregator node is elected for complex processing. The election of a node is based on the energy reserves and computational capability of that node. The algorithm has two components. The first computes the signaling overhead associated with the election process of the single node; the node with the least overhead will be the winner. The winner node broadcasts a message with its ID that will be stored in the node's registry. The second component of the algorithm finds a spanning tree rooted at the winner node. The building of the spanning tree follows a procedure similar to Kruskal's algorithms outlined in Sohrabi and Pottie [11]. By the end of the SWE process, a minimum-hop spanning tree will completely cover the network.

In the multiple winner algorithm (MWE), a simple extension to the SWE is proposed. When all nodes are sources and send their data to the central aggregator node, a large amount of energy will be consumed, so this process has a high cost. One way to lower the energy cost is to limit the number of sources that can send data to the central aggregator node. Instead of keeping record of only the best candidate node (master aggregator node), each node will keep a record of up to n nodes of those candidates. At the end of the MWE process, each sensor in the network has a set of minimum-energy paths to each source node (SN). After that, the SWE is used to find the node that yields the minimum energy consumption. This node can then serve as the central node for the coherent processing. In general, the MWE process has longer delay, higher overhead, and lower scalability than that for noncoherent processing networks.

3.2.2 Hierarchical Routing

Hierarchical or cluster-based routing, originally proposed in wireline networks, comprises well-known techniques with special advantages related to scalability and efficient communication. As such, the concept of hierarchical routing is also utilized to perform energy-efficient routing in WSNs. In a hierarchical architecture, higher energy nodes can be used to process and send the information while low energy nodes can be used to perform the sensing in the proximity of the target. This means that creation of clusters and assigning special tasks to cluster heads can greatly contribute to overall system scalability, lifetime, and energy efficiency.

3.2.2.1 LEACH Protocol

Heinzelman et al. [1] introduced a hierarchical clustering algorithm for sensor networks called low energy adaptive clustering hierarchy (LEACH). LEACH is a cluster-based protocol that includes distributed cluster formation. The authors allowed for a randomized rotation of the cluster head's role in the objective of reducing energy consumption (i.e., extending network lifetime) and to distribute the energy load evenly among the sensors in the network. LEACH uses localized coordination to enable scalability and robustness for dynamic networks and incorporates data fusion into the routing protocol in order to reduce the amount of information that must be transmitted to the base station. The authors also made use of a TDMA/CDMA MAC to reduce inter- and intracluster collisions.

Because data collection is centralized and performed periodically, this protocol is most appropriate when constant monitoring by the sensor network is needed. A user may not need all the data immediately. Thus, periodic data transmissions, which may drain the limited energy of the sensor nodes, are unnecessary. The authors of LEACH introduced adaptive clustering, i.e., reclustering after a given interval with a randomized rotation of the energy-constrained cluster head so that energy dissipation in the sensor network is uniform. They also found, based on their simulation model, that only 5% of the nodes need to act as cluster heads.

The operation of LEACH is separated into two phases: the setup phase and the steady state phase. In the setup phase, the clusters are organized and cluster heads are selected. In the steady state phase, the actual data transfer to the base station takes place. The duration of the steady state phase is longer than the duration of the setup phase in order to minimize overhead.

During the setup phase, a predetermined fraction of nodes, p, elect themselves as cluster heads as follows. A sensor node chooses a random number, r, between 0 and 1. If this random number is less than a threshold value, $T(n)$, the node becomes a cluster head for the current round. The threshold value is calculated based on an equation that incorporates the desired percentage to become a cluster head, the current round, and the set of nodes not selected as a cluster head in the last $(1/P)$ rounds, denoted by G. This is given by:

$$T(n) = \frac{p}{1 - p(r \ mod \ (1/p))} \quad if \ n \in G$$

where G is the set of nodes that

After the cluster heads have been elected, they broadcast an advertisement message to the rest of the nodes in the network that they are the new cluster heads. Upon receiving this advertisement, all the noncluster head nodes decide on the cluster to which they want to belong, based on the signal strength of the advertisement. The noncluster head nodes inform the appropriate cluster heads that they will be members of the cluster. Figure 3.8 shows a flowchart of the cluster head election procedure.

After receiving all the messages from the nodes that would like to be included in the cluster and based on the number of nodes in the cluster, the cluster head node creates a TDMA schedule and assigns each node a time slot when it can transmit. This schedule is broadcast to all the nodes in the cluster. During the steady state phase, the sensor nodes can begin sensing and transmitting data to the cluster heads. The cluster head node, after receiving all the data, aggregates them before sending them to the base station. After a certain time, which is determined *a priori*, the network goes back into the setup phase

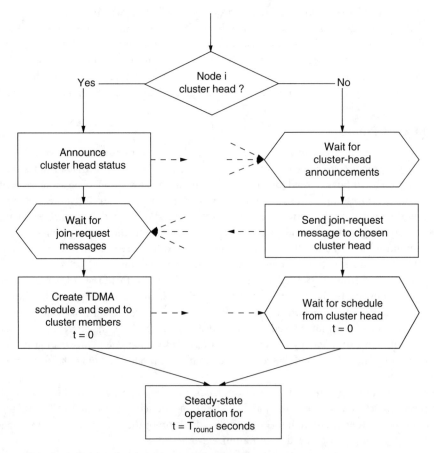

FIGURE 3.8 Flowchart of cluster head election in LEACH protocol.

again and enters another round of selecting new cluster heads. Each cluster communicates using different CDMA codes to reduce interference from nodes belonging to other clusters.

Although LEACH is able to increase the network lifetime, a number of issues about the assumptions used in this protocol remain. LEACH assumes that all nodes can transmit with enough power to reach the base station if needed and that each node has computational power to support different MAC protocols. It also assumes that nodes always have data to send, and nodes located near each other have correlated data. It is not obvious how the number of the predetermined cluster heads (p) is going to be uniformly distributed through the network. Because it is possible that the elected cluster heads will be concentrated in one part of the network, some nodes will not have any cluster heads in their vicinity. Finally, the protocol assumes that all nodes begin with the same amount of energy capacity, supposing that a cluster head removes approximately the same amount of energy for each node. The protocol should be extended to account for nonuniform energy nodes, i.e., use energy-based threshold.

Heinzelman and coworkers proposed an extension to LEACH — LEACH with negotiation [7]. The main theme of the proposed extension is that high-level negotiation using metadata descriptors (as in the SPIN protocol discussed in Section 3.2.3) precede data transfers. This ensures that only data that provide new information are transmitted to the cluster heads before being transmitted to the base station.

3.2.2.2 Power-Efficient Gathering in Sensor Information Systems (PEGASIS)

In Lindsey and Raghavendra [12], an enhancement over the LEACH protocol was proposed. This protocol, called power-efficient gathering in sensor information systems (PEGASIS), is a near optimal chain-based protocol. The basic idea of the protocol is that, in order to extend network lifetime, nodes need only communicate with their closest neighbors and take turns in communicating with the base station. When the round of all nodes communicating with the base station ends, a new round will start and so on. This reduces the power required to transmit data per round because the power draining is spread uniformly over all nodes. Thus, PEGASIS has two main objectives: (1) to increase the lifetime of each node by using collaborative techniques and thus increase network lifetime; and (2) to allow only local coordination between nodes that are close together so that the bandwidth consumed in communication is reduced.

To locate the closest neighbor node, each node uses signal strength to measure the distance to all neighboring nodes and then adjusts the strength so that only one node can be heard. The chain in PEGASIS will consist of nodes closest to each other that form a path to the base station. The aggregated form of the data will be sent to the base station by any node in the chain and the nodes in the chain will take turns sending to the base station. The authors show through simulation that PEGASIS is able to increase the lifetime of the network to twice the lifetime of the network under the LEACH protocol.

However, PEGASIS uses assumptions that may not always be realistic. First, PEGASIS assumes that each sensor node is able to communicate with the base station directly. In practical cases, sensor nodes use multihop communication to reach the base station. Second, it assumes that all nodes maintain a complete database about the location of all other nodes in the network, but the method by which the node locations are obtained is not outlined. Third, it assumes that all sensor nodes have the same level of energy and are likely to die at the same time. Fourth, although in most scenarios sensors will be fixed or immobile as assumed in PEGASIS, some sensors may be allowed to move and thus affect the protocol functions.

3.2.2.3 Threshold-Sensitive Energy-Efficient Protocols (TEEN and APTEEN)

Two hierarchical routing protocols called TEEN (threshold-sensitive energy-efficient sensor network) and APTEEN (adaptive periodic threshold-sensitive energy-efficient sensor network) have been proposed by Manjeshwar and Agarwal [8,9] for time-critical applications. In TEEN, sensor nodes sense the medium continuously, but the data transmission is done less frequently. A cluster head sensor sends its members a hard threshold, which is the threshold value of the sensed attribute, and a soft threshold, which is a small change in the value of the sensed attribute that triggers the node to switch on its transmitter and

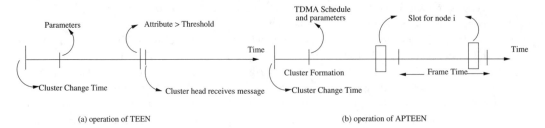

(a) operation of TEEN   (b) operation of APTEEN

FIGURE 3.9 Time line for the operation of (a) TEEN and (b) APTEEN.

transmit. Thus, the hard threshold tries to reduce the number of transmissions by allowing the nodes to transmit only when the sensed attribute is in the range of interest.

The soft threshold further reduces the number of transmissions that might have otherwise occurred when little or no change occurs in the sensed attribute. A smaller value of the soft threshold gives a more accurate picture of the network, at the expense of increased energy consumption. Thus, the user can control the trade-off between energy efficiency and data accuracy. When cluster heads are to change (see Figure 3.9), new values for the preceding parameters are broadcast. The main drawback of this scheme is that, if the thresholds are not received, the nodes will never communicate and the user will not get any data from the network.

The nodes sense their environment continuously. The first time a parameter from the attribute set reaches its hard threshold value, the node switches on its transmitter and sends the sensed data. The sensed value is stored in an internal variable, called sensed value (SV). The nodes will transmit data in the current cluster period only when the following conditions are true: (1) the current value of the sensed attribute is greater than the hard threshold ; and (2) the current value of the sensed attribute differs from SV by an amount equal to or greater than the soft threshold.

Important features of TEEN include its suitability for time-critical sensing applications. Also, because message transmission consumes more energy than data sensing, the energy consumption in this scheme is less than the proactive networks. The soft threshold can be varied. At every cluster change time, the parameters are broadcast afresh, so the user can change them as required. The main drawback is that if the thresholds are not reached, the nodes will never communicate.

APTEEN, on the other hand, is a hybrid protocol that changes the periodicity or threshold values used in the TEEN protocol according to user needs and type of the application. In APTEEN, the cluster heads broadcast the following parameters:

- *Attributes* (A) is a set of physical parameters about which the user is interested in obtaining information.
- *Thresholds* consist of the hard threshold (HT) and the soft threshold (ST).
- *Schedule* is a TDMA schedule that assigns a slot to each node.
- *Count time* (CT) is the maximum time period between two successive reports sent by a node.

The node senses the environment continuously and only nodes that sense a data value at or beyond the hard threshold transmit. Once a node senses a value beyond HT, it transmits data only when the value of that attribute changes by an amount equal to or greater than the ST. If a node does not send data for a time period equal to the count time, it is forced to sense and retransmit the data. A TDMA schedule is used and each node in the cluster is assigned a transmission slot. Thus, APTEEN uses a modified TDMA schedule to implement the hybrid network. The main features of the APTEEN scheme include: (1) combining proactive and reactive policies; (2) offering a lot of flexibility by allowing the user to set the CT interval; and (3) controlling threshold values for the energy consumption by changing the CT as well as the threshold values. The main drawback of the scheme is the additional complexity required to implement the threshold functions and the CT. However, the authors of these two protocols showed through simulation that both protocols perform better than LEACH.

3.2.2.4 Small Minimum Energy Communication Network (SMECN)

Rodoplu and Meng [16] have proposed a protocol that computes an energy-efficient subnetwork, namely, the minimum energy communication network (MECN), for a certain sensor network. A new algorithm called small MECN (SMECN) to provide such a subnetwork has been proposed by Li and Halpern [17]. The subnetwork (i.e., subgraph G') constructed by SMECN is smaller than the one constructed by MECN if the broadcast region is circular around the broadcasting node for a given power setting. Subgraph G' of graph G, which represents the sensor network, minimizes the energy usage satisfying the following conditions: (1) the number of edges in G' is less than in G while containing all nodes in G; and (2) the energy required to transmit data from a node to all its neighbors in subgraph G' is less than the energy required to transmit to all its neighbors in graph G.

Assuming that $r = (u, u_1,\ldots, u_{k-1}, v)$ is a path between u and v, the total power consumption of one path like r is given by:

$$C(r) = \sum_{i=0}^{k-1} (p(u_i, u_{i+1}) + c)$$

where $u = u_0$; $v = u_k$; the power required to transmit data under this protocol is

$$p(u, v) = td(u, v)^n$$

for some appropriate constant t; n is the path-loss exponent of outdoor radio propagation models $n \geq 2$ and $d(u,v)$ is the distance between u and v. A reception at the receiver takes power c.

The subnetwork computed by SMECN helps to send messages on minimum-energy paths. However, the proposed algorithm is local in the sense that it does not actually find the minimum-energy path; it just constructs a subnetwork in which the path is guaranteed to exist. Moreover, the subnetwork constructed by SMECN makes it more likely that the path used is one that requires less energy consumption.

3.2.2.5 Fixed-Size Cluster Routing

Xu and colleagues [19] have proposed a geography informed routing protocol for ad hoc networks. The network area is first divided into fixed zones; inside each zone, nodes collaborate with each other to play different roles. For example, nodes will elect one sensor node to stay awake for a certain period of time and then they go to sleep. Each sensor node is positioned randomly in a two-dimensional plane. When a sensor transmits a packet with power for a distance r, the signal will be strong enough for other sensors to hear it within the Euclidean distance r from the sensor that originates the packet. In other words, to cover a range of r, the sensor that originates the signal must transmit with enough power to cover that range.

Figure 3.10 gives an example of fixed zoning that can be used in sensor networks similar to the one proposed by Xu et al., but with an extension. The extension is to use two zones to receive signals instead of one. After the range r, the power signal starts to attenuate (i.e., fade out), so a sensor in the second zone, called the *border* zone, may or may not hear the signal depending on the signal strength. Therefore, a sensor within the *guaranteed* zone, i.e., within the distance r, is guaranteed to receive the signal, while a sensor in the border zone may or may not receive the packet. Figure 3.10 shows this situation.

Xu and colleagues' fixed clusters [19] are selected to be equal and square. The selection of the square size depends on the required transmitting power and the communication direction. One node in each cluster, called the cluster head, is elected periodically. Vertical and horizontal communication is guaranteed if the signal travels a distance of $a = \dfrac{r}{\sqrt{5}}$, chosen so that any two sensor nodes in adjacent vertical or horizontal clusters can communicate directly in the guaranteed zone. For a node in the *border* zone

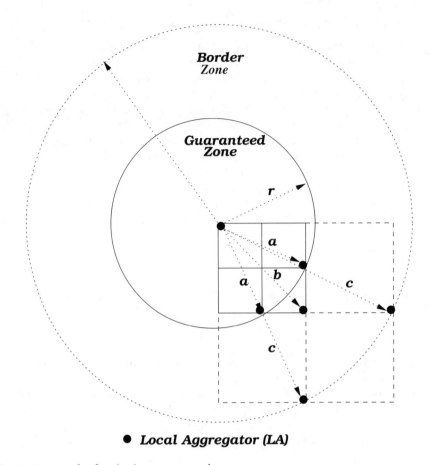

● **Local Aggregator (LA)**

FIGURE 3.10 An example of zoning in sensor networks.

to receive the transmitted packet, the signal must travel a distance of $c = \dfrac{r}{2\sqrt{5}}$. Note also that for a

diagonal communication to happen, the signal must span a distance of $b = \dfrac{r}{2\sqrt{2}}$. A cluster head is

responsible for receiving raw data from other nodes in its cluster. The role of cluster head is rotated to distribute the energy draining role evenly around the network.

3.2.2.6 Virtual Grid Architecture Routing

An energy-efficient routing paradigm proposed by [26] is based on the concept of data aggregation and in-network processing. The data aggregation is performed at two levels: local and then global. A reasonable approach for WSNs is to arrange nodes in a fixed topology due to the node stationarity or extremely low mobility. Fixed, equal, adjacent, and nonoverlapping clusters with regular shapes are selected to obtain a fixed rectilinear virtual topology. Inside each zone, a node is optimally selected to act as cluster head. The set of cluster heads, also called local aggregators (LAs), performs the local aggregation. Several heuristics were formulated to allocate a subset of the cluster heads, called the master aggregators (MAs), in order to perform near optimal global data aggregation so that the total routing cost from the source nodes to the base station is minimized.

Figure 3.11 illustrates an example of fixed zoning and the resulting virtual grid architecture (VGA) used to perform two level data aggregation. Note that the location of the base station is not necessarily at the extreme corner of the grid, but rather can be located at an arbitrary place.

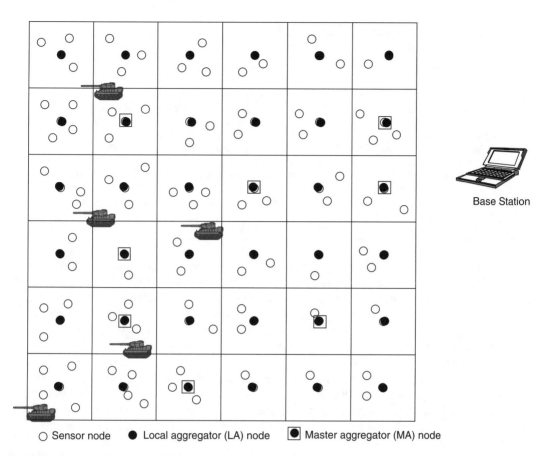

○ Sensor node ● Local aggregator (LA) node ▣ Master aggregator (MA) node

FIGURE 3.11 Regular shape tessellation applied to the network area. In each zone, a cluster head is selected for local aggregation. A subset of those cluster heads, called master nodes, are optimally selected to perform global aggregation.

All heuristics in Reference 26 start with the first node in the VGA architecture and proceed sequentially the whole topology left to right and then right to left in a top-down fashion. Although finding the optimal routes from the source nodes to the base station by using the set of MAs is an NP-complete problem, Al-Karaki and Kamal's developed dynamic program [26] is able to find the optimal values most of the time.

3.2.2.7 Hierarchical Power-Aware Routing

Li and coworkers [20] have proposed a hierarchical power-aware routing protocol that divides the network into groups of sensors. Each group of sensors in geographic proximity is clustered together as a zone and each zone is treated as an entity. To perform routing, each zone is allowed to decide how it will route a message hierarchically across the other zones.

Messages are routed along the path with the maximal–minimal of the remaining power, called the max–min path. The motivation is that using nodes with high residual power may be expensive compared to the path with the minimal power consumption. An approximation algorithm, called the *max–min zPmin* algorithm, combines the benefits of selecting the path with the minimum power consumption and the path that maximizes the minimal residual power in the nodes of the network. The algorithm finds the path with the least power consumption, P_{min}, by using the Dijkstra algorithm.

Another algorithm, called zone-based routing, that relies on *max–min zPmin* and is scalable for large scale networks has also been proposed in Reference 20. Zone-base routing is a hierarchical approach in which the area covered by the (sensor) network is divided into a small number of zones. To send a

TABLE 3.1   Hierarchical vs. Flat Topology Routing

Hierarchical Routing	Flat Routing
Reservation-based scheduling	Contention-based scheduling
Collisions avoided	Collision overhead present
Reduced duty cycle due to periodic sleeping	Variable duty cycle by controlling sleep time of nodes
Data aggregation by cluster head	Node on multihop path aggregates incoming data from neighbors
Simple but nonoptimal routing	Routing is complex but optimal
Requires global and local synchronization	Links formed on the fly without synchronization
Overhead of cluster formation throughout the network	Routes formed only in regions with data for transmission
Lower latency because multiple hops network formed by cluster heads always available	Latency in waking up intermediate nodes and setting up multipath
Energy dissipation is uniform	Energy dissipation depends on traffic patterns
Energy dissipation cannot be controlled	Energy dissipation adapts to traffic pattern
Fair channel allocation	Fairness not guaranteed

message across the entire area, a global path from zone to zone is found. The sensors in a zone autonomously direct local routing and participate in estimating the zone power level. Each message is routed across the zones using information about the zone power estimates. A global controller for message routing, which may be the node with the highest power, is assigned the role of managing the zones. If the network can be divided into a relatively small number of zones, the scale for the global routing algorithm is reduced. The global information required to send each message across is summarized by the power level estimate of each zone.

A zone graph was used to represent connected neighboring zone vertices if the current zone can go to the next neighboring zone in that direction. Each zone vertex has a power level of 1. Each zone direction vertex is labeled by its estimated power level, computed by a procedure that is a modified Bellman–Ford algorithm. Moreover, two algorithms were outlined for local and global path selection using the zone graph.

The flat and hierarchical protocols are different in many aspects. Table 3.1 outlines the major differences between the two routing approaches.

3.2.3 Adaptive Routing

Heinzelman et al. [3] and Kulik et al. [6] proposed a family of adaptive protocols, called sensor protocols for information via negotiation (SPIN). These protocols disseminate all the information at each node to every node in the network, assuming that all nodes in the network are potential base stations. This enables a user to query any node and get the required information immediately. These protocols make use of the property that nearby nodes have similar data and thus distribute only data that the other nodes do not have.

The SPIN family of protocols uses data negotiation and resource-adaptive algorithms. Nodes running SPIN assign a high-level name to describe their collected data (called metadata) completely and perform metadata negotiations before any data are transmitted. This assures that no redundant data are sent throughout the network. The format of the metadata is application specific and is not specified in SPIN. For example, sensors might use their unique IDs to report metadata if they cover a certain known region. In addition, SPIN has access to the current energy level of the node and adapts the protocol it is running based on how much energy is remaining. These protocols work in a time-driven fashion and distribute the information over the network, even when a user does not request any data.

The SPIN family is designed to address the deficiencies of classic flooding by negotiation and resource adaptation. This family of protocols is designed based on the idea that sensor nodes operate more efficiently and conserve more energy by sending data that describe the sensor data instead of sending all the data; for example, image and sensor nodes must monitor the changes in their energy resources.

SPIN protocols are motivated by the observation that conventional protocols like flooding or gossiping waste energy and bandwidth by sending extra and unnecessary copies of data by sensors covering overlapping areas. Sensor nodes use three types of messages — ADV, REQ, and DATA — to communicate. ADV advertises new data, REQ requests data, and DATA is the actual message. The protocol starts when a SPIN node obtains new data that it is willing to share. It does so by broadcasting an ADV message containing metadata. If a neighbor is interested in the data, it sends a REQ message for the DATA and the DATA is sent to this neighbor node. The neighbor sensor node then repeats this process with its neighbors. As a result, the entire sensor area will receive a copy.

The SPIN family of protocols includes two protocols, namely, SPIN-1 and SPIN-2, which incorporate negotiation before transmitting data in order to eliminate implosion and overlap by ensuring that only useful information will be transferred. Also, each node has its own resource manager, which keeps track of resource consumption, and is polled by the nodes before data transmission. The SPIN-1 protocol is a three-stage protocol, as described earlier. An extension to SPIN-1 is SPIN-2, which incorporates a threshold-based resource awareness mechanism in addition to negotiation. When energy in the nodes is abundant, SPIN-2 communicates using the three-stage protocol of SPIN-1.

However, when the energy in a node starts approaching a low energy threshold, it reduces its participation in the protocol, i.e., it participates only when it believes that it can complete all the other stages of the protocol without going below the low-energy threshold. This approach does not prevent a node from receiving, and therefore spending, energy on ADV, or REQ messages below its low-energy threshold. It does, however, prevent the node from ever handling a DATA message below this threshold.

In conclusion, SPIN-l and SPIN-2 are simple protocols that efficiently disseminate data while maintaining no per-neighbor state. These protocols are well suited for an environment in which the sensors are mobile because they base their forwarding decisions on local neighborhood information. Other protocols of the SPIN family are:

- SPIN-BC. This protocol is designed for broadcast channels. All nodes within hearing range of a sensor node will get the message. However, nodes must wait for transmission if the channel is busy. Also, nodes do not immediately send out REQ message when they hear the ADV message. Instead, each node sets a random timer and when this timer expires, the node sends the REQ message. If, waiting for their timers to expire, other nodes are able to hear this message, they will stop their timers. This prevents sending redundant copies of the same request.
- SPIN-PP. If two nodes can communicate with each other without incurring interference from other neighboring nodes, this protocol will be used. It is designed for a point-to-point communication, i.e., hop-by-hop routing, and assumes that energy is not a major constraint and that packets are never lost. Figure 3.12 shows an example of the operation of this protocol. A node will send an ADV message to advertise that it has a message to send. All nodes in the neighborhood that hear the message, if interested, will express this interest by sending REQ messages. Upon receiving the REQ message, the announcing node will send the data to the interested nodes. Once those nodes have the information, they become an information announcer and send an ADV message to their neighbors. If their neighbors are interested, they send an REQ message and the process repeats.
- SPIN-EC. This protocol works similarly to SPIN-PP, but with an energy heuristic added to it. A node will participate in the protocol if the node is able to complete all stages of the protocol without its energy dropping below a certain threshold. The energy threshold is a system parameter.
- SPIN-RL. In SPIN-PP, it is assumed that packets are not lost. When a channel is lossy, this protocol cannot be used. Instead, another protocol called SPIN-RL, in which two adjustments are added to the SPIN-PP protocol to account for the lossy channel, is used. First, each node keeps track of all ADV messages it receives. It may also ask for data to be resent if it did not get them within a specified amount of time. Second, in order to fine tune the rate of resending data, nodes will limit the frequency of this activity by having each node wait for a certain predetermined time before replying to the same REQ messages again. This procedure guarantees that data will be resent only after making sure that the reply to the previous REQ message failed.

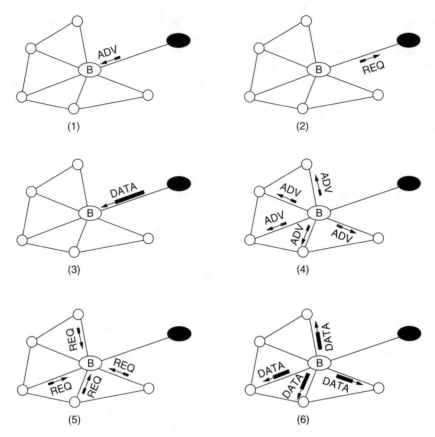

FIGURE 3.12 SPIN-PP: three-way handshake in SPIN protocol. Steps 1 through 6 show the three messages (ADV, REQ, and DATA) used in the handshaking process.

Table 3.2 compares SPIN, LEACH, and the directed diffusion routing techniques according to different parameters. The table indicates that directed diffusion shows a promising approach for energy-efficient routing in WSNS due to the use of in-network processing.

3.2.4   Multipath Routing

The resilience of a protocol is measured by the likelihood that an alternate path exists between a source and a sink when the primary path fails. This can be increased by maintaining multiple paths between the source and the sink at the expense of increased energy consumption, and keeping these alternate paths alive by sending periodic messages. Thus, the resilience of the network should be increased while keeping the maintenance overhead of these paths low. This subsection discusses routing protocols that use multiple paths rather than a single path in order to enhance network performance.

TABLE 3.2   Comparison among SPIN, LEACH, and Directed Diffusion

	SPIN	LEACH	Directed Diffusion
Optimal route	No	No	Yes
Network lifetime	Good	Very good	Good
Resource awareness	Yes	Yes	Yes
Use of metadata	Yes	No	Yes

Ganesan and coworkers [22] have proposed an energy-efficient multipath routing protocol that uses braided multipaths instead of completely disjoint multipaths so as to keep the cost of maintenance low. The costs of such alternate paths are also comparable to the primary path because they tend to be much closer to the primary path. Chang and Tassiulas [23] proposed an algorithm to route data through a path whose nodes have the largest residual energy. The path is changed whenever a better path is discovered. The primary path will be used until its energy falls below the energy of the backup path at which the backup path is used. In this way, the nodes in the primary path will not deplete their energy resources through continual use of the same route, thus achieving longer life. The path-switching cost was not quantified in the paper.

Rahul and Rabaey [24] have proposed the use of a set of suboptimal paths occasionally to increase the lifetime of the network. These paths are chosen by means of a probability that depends on how low the energy consumption of each path is.

Because the path with the largest residual energy when used to route data in a network may be very energy expensive too, a trade-off takes place between minimizing the total power consumed and the residual energy of the network. Li and colleagues [20] proposed an algorithm in which the residual energy of the route is relaxed a bit in order to select a more energy-efficient path. The operation of the algorithm is explained in Subsection 3.2.2.7.

3.2.5 Query-Based Routing

In this kind of routing, the destination nodes propagate a query for data (sensing task) from a node through the network and a node having these data sends data that match the query back to the node, which initiates the query. Usually these queries are described in natural language, or in high-level query languages. For example, client C1 may submit a query to node N1 and ask, "Are there moving vehicles in battle space region 1?"

All the nodes have tables consisting of the sensing task queries received, and hence they send data that match these queries when they receive them. Directed diffusion (described in Subsection 3.2.1.2) is an example of this type of routing. In directed diffusion, the sink node sends out interest messages to sensors. As the interest is propagated throughout the sensor network, the gradients from the source back to the sink are set up. When the source has data for the interest, the source sends the data along the interest's gradient path. To lower energy consumption, data aggregation (e.g., duplicate suppression) is performed en route.

3.2.6 Negotiation-Based Protocols

These protocols use high-level data descriptors in order to eliminate redundant data transmissions through negotiation. Communication decisions are also taken based on the resources available to them.

The SPIN family protocols discussed in Section 3.2.3 are an example of negotiation-based routing protocols. The motivation is that the use of flooding to disseminate data will produce implosion and overlap among the sent data and thus nodes will receive duplicate copies of the same data. This operation consumes more energy and more processing by sending the same data by different sensors. The SPIN protocols are designed to disseminate the data of one sensor to all other sensors assuming these sensors are potential base stations. Therefore, the main idea of negotiation-based routing in WSNs is to suppress duplicate information and prevent redundant data from being sent to the next sensor or the base station by conducting a series of negotiation messages before the real data transmission begins.

3.3 Routing in WSNs: Future Directions

The future vision of WSNs is to embed numerous distributed devices to monitor and interact with physical world phenomena and to exploit spatially and temporally dense sensing and actuation capabilities of those sensor networks. These nodes coordinate among themselves to create a network that performs higher level tasks.

Although extensive efforts have been exerted so far on the routing problem in WSNs, some challenges still confront effective solutions of the routing problem. First, there is a tight coupling between sensor nodes and the physical world. Sensors are embedded in *unattended* places or systems. This is different from traditional Internet, PDA, and mobility applications that interface primarily and directly with human users. Second, sensors are characterized by a small footprint and, as such, nodes present stringent energy constraints because they are living with small, finite, energy sources. This is also different from traditional fixed but reusable resources. Third, communications is primary consumer of energy in this environment in which sending a bit over 10 or 100 m consumes as much energy as thousands to millions of operations (known as R^4 signal energy drop-off) [27].

Future trends in routing techniques in WSNs focus on different directions, but all share the common objective of prolonging network lifetime. Some of these directions include:

- *Exploit redundancy.* Typically, a large number of sensor nodes are implanted inside or beside the phenomenon. Because sensor nodes are prone to failure, fault tolerance techniques come into the picture to keep the network operating and performing its tasks. Routing techniques that explicitly employ fault tolerance techniques in an efficient manner are still under investigation.
- *Tiered architectures (mix of form/energy factors).* Hierarchical routing is an old technique to enhance scalability and efficiency of the routing protocol. However, novel techniques to network clustering to maximize the network lifetime are also a hot area of research in WSNs.
- *Exploit spatial diversity and density of sensor/actuator nodes.* Nodes will span a network area that might be large enough to provide spatial communication between sensor nodes. Achieving energy efficient communication in this densely populated environment deserves further investigation. The dense deployment of sensor nodes should allow the network to adapt to unpredictable environments.
- *Achieve desired global behavior with adaptive localized algorithms.* That is, do not rely on global interaction or information. However, in a dynamic environment, this is hard to model.
- *Leverage data processing inside the network and exploit computation near data sources to reduce communication.* That is, perform in-network distributed processing. WSNs are organized around naming data, not node identities. Because a large collection of distributed elements is present, localized algorithms that achieve system-wide properties in terms of local processing of data before being sent to the destination are still needed. Nodes in the network will store named data and make them available for processing. The need is great to create efficient processing points in the network, e.g., duplicate suppression, aggregation, correlation of data. How to find those points efficiently and optimally is still an open research issue.
- *Time and location synchronization.* Energy-efficient techniques for associating time and spatial coordinates with data to support collaborative processing are also required.
- *Self-configuration and reconfiguration.* These are essential to the lifetime of unattended systems in dynamic, constrained-energy environments and important for keeping the network up and running. As nodes die and leave the network, update and reconfiguration mechanisms should take place. An important feature in every routing protocol is to adapt to topology changes very quickly and to maintain the network functions.

3.4 Conclusions

Routing in sensor networks is a new area of research, with a limited but rapidly growing set of research results. This chapter offered a comprehensive overview of routing techniques in wireless sensor networks that have been presented in the literature. They have the common objective of trying to extend the lifetime of the sensor network.

Overall, the routing techniques are classified based on the network structure into three categories: flat, hierarchical, and adaptive routing. Furthermore, these protocols can be classified into multipath-based, query-based, or negotiation-based routing techniques depending on the protocol operation. Design trade-offs between energy and communication overhead savings in some of the routing paradigm have been highlighted, as well as advantages and disadvantages of each routing technique. Although many of these routing techniques look promising, many challenges in the sensor networks still need to be solved; this chapter highlighted those challenges and pinpointed future research directions in this regard.

References

1. W. Heinzelman, A. Chandrakasan, and H. Balakrishnan, Energy-efficient communication protocol for wireless microsensor networks, *Proc. 33rd Hawaii Int. Conf. Syst. Sci.* (HICSS '00), January 2000.
2. C. Intanagonwiwat, R. Govindan, and D. Estrin, Directed diffusion for wireless sensor networks, *IEEE/ACM Trans. Networking*, 11(1), 2–16, 2003.
3. W. Heinzelman, J. Kulik, and H. Balakrishnan, Adaptive protocols for information dissemination in wireless sensor networks, *Proc. 5th ACM/IEEE Mobicom Conf. (MobiCom'99)*, Seattle, WA, August, 1999. 174–185.
4. I. Akyildiz, W. Su, Y. Sankarasubramaniam, and E. Cayirci, A survey on sensor networks, *IEEE Commun. Mag.*, 40(8), 102–114, August 2002.
5. A. Perrig, R. Szewzyk, J.D. Tygar, V. Wen, and D. E. Culler, SPINS: security protocols for sensor networks. *Wireless Networks*, 8, 521–534, 2000.
6. J. Kulik, W.R. Heinzelman, and H. Balakrishnan, Negotiation-based protocols for disseminating information in wireless sensor networks, *Wireless Networks*, 8, 169–185, 2002.
7. W.R. Heinzelman, A. Chandrakasan, and H. Balakrishnan, Energy-efficient communication protocol for wireless microsensor networks, *Proc. 33rd Int. Conf. Syst. Sci.*, (HICSS'00), January 2000, 1–10.
8. A. Manjeshwar and D.P. Agarwal, TEEN: a routing protocol for enhanced efficiency in wireless sensor networks, in *1st Int. Workshop Parallel Distributed Computing Issues Wireless Networks Mobile Computing*, April 2001.
9. A. Manjeshwar and D.P. Agarwal, APTEEN: a hybrid protocol for efficient routing and comprehensive information retrieval in wireless sensor networks, *Parallel Distributed Process. Symp., Proc. Int.*, IPDPS 2002, 195–202.
10. D. Ganesan, R. Govindan, S. Shenker, and D. Estrin, Highly-resilient, energy-efficient multipath routing in wireless sensor networks, *ACM SIGMOBILE Mobile Computing Commun. Rev.*, 5(4), 10–24, October 2001.
11. K. Sohrabi and J. Pottie, Protocols for self-organization of a wireless sensor network, *IEEE Personal Commun.* 7(5), 16–27, 2000.
12. S. Lindsey and C. Raghavendra, PEGASIS: power-efficient gathering in sensor information systems, *Int. Conf. Communication Protocols*, 149–155, 2001.
13. F. Ye, A. Chen, S. Liu, and L. Zhang, A scalable solution to minimum cost forwarding in large sensor networks, *Proc. 10th Int. Conf. Computer Commun. Networks (ICCCN)*, 304–309, 2001.
14. N. Bulusu, J. Heidemann, and D. Estrin, GPS-less low cost outdoor localization for very small devices, Technical report 00-729, Computer Science Department, University of Southern California, Apr. 2000.
15. A. Savvides, C.-C. Han, and M. Srivastava, Dynamic fine-grained localization in Ad-Hoc networks of sensors, *Proc. 7th ACM Annu. Int. Conf. Mobile Computing Networking (MobiCom)*, July 2001, 166–179.
16. V. Rodoplu and T.H. Meng, Minimum energy mobile wireless networks, IEEE JSAC, 17(8), Aug. 1999, 1333–1344.
17. L. Li and J.Y. Halpern, Minimum-energy mobile wireless networks revisited, *ICC '01*, Helsinki, Finland, 67–78, June 2001.
18. S. Hedetniemi, S. Hedetniemi, and A. Liestman, A survey of gossiping and broadcasting in communication networks, *Networks*, 18, 1988.

19. Y. Xu, J. Heidemann, D. Estrin, Geography-informed energy conservation for ad-hoc routing, *IEEE/ACM MobiCom*, Rome, 70–84, July 16–21, 2001.

20. Q. Li, J. Aslam, and D. Rus, Hierarchical power-aware routing in sensor networks, in *Proc. DIMACS Workshop Pervasive Networking*, May, 2001.

21. D. Braginsky and D. Estrin, Rumor routing algorithm for sensor networks, ACM First Workshop on Sensor Networks and Applications (WSNA), 2002.

22. D. Ganesan, R. Govindan, S. Shenker, and D. Estrin, Highly resilient, energy-efficient multipath routing in wireless sensor networks, *ACM Mobile Computing Commun. Rev.*, 5(4), October 2001.

23. J.-H. Chang and L. Tassiulas, Maximum lifetime routing in wireless sensor networks, *Proc. Adv. Telecommun. Inf. Distribution Res. Program (ATIRP2000)*, College Park, MD, Mar. 2000.

24. C. Rahul and J. Rabaey, Energy-aware routing for low energy ad hoc sensor networks, *IEEE Wireless Commun. Networking Conf. (WCNC)*, March 17–21, 2002, Orlando, FL.

25. W. Heinzelman, J. Kulik, and H. Balakrishnan, Adaptive protocols for information dissemination in wireless sensor networks, *Proc. 5th Annu. ACM/IEEE Int. Conf. Mobile Computing Networking*, August 1999.

26. J. Al-Karaki and A. Kamal, On the optimal data aggregation and in-network processing based routing in wireless sensor networks, technical report, Iowa State University, 2003.

27. D. Goodman, *Wireless Personal Communications Systems*. Reading, MA: Addison–Wesley, Reading, MA, 1997.

4

Overview of Communication Protocols for Sensor Networks

Weilian Su
Georgia Institute of Technology

Erdal Cayirci
Istanbul Technical University

Özgür B. Akan
Georgia Institute of Technology

4.1 Introduction

As the technology for wireless communications advances and the cost of manufacturing a sensor node continues to decrease, a low-cost but yet powerful sensor network may be deployed for various applications that can be envisioned for daily life. Although each sensor node may seem to be much less capable than a traditional stationary sensor, a collective effort of the sensor nodes may provide sensing capabilities in space and time that surpass the stationary sensor.

The communication protocols for sensor networks may leverage the capabilities of collective efforts to provide users with specialized applications. These protocols may fuse, extract, or aggregate data from the sensor field. In addition, they may self-organize the sensor nodes into clusters to complete a task or overcome certain obstacles, e.g., hills. In essence, sensor networks may provide end users with intelligence and details that traditional stationary sensors may not be able to do.

Although the sensor nodes communicate through the wireless medium, protocols and algorithms proposed for traditional wireless ad hoc networks may not be well suited for sensor networks. As previously explained, sensor networks are application specific, and the sensor nodes work collaboratively together. In addition, the sensor nodes are very energy constrained compared to traditional wireless ad hoc devices. The differences between sensor networks and ad hoc networks [29] are:

- The number of sensor nodes in a sensor network can be several orders of magnitude higher than the nodes in an ad hoc network.

- Sensor nodes are densely deployed.
- Sensor nodes are prone to failures.
- The topology of a sensor network changes very frequently.
- Sensor nodes mainly use a broadcast communication paradigm whereas most ad hoc networks are based on point-to-point communications.
- Sensor nodes are limited in power, computational capacities, and memory.
- Sensor nodes may not have global identification (ID) because of the large amount of overhead and large number of sensor nodes.
- Sensor networks are deployed with a specific sensing application in mind; ad hoc networks are mostly constructed for communication purposes.

With these differences, the design of communication protocols for sensor networks requires specific attention. Some of the potential applications as well as some application layer protocols for sensor networks are presented in Section 4.2. Next, because many of the communication protocols require the knowledge of location and time in order to function properly, localization and time synchronization protocols are described in Section 4.3 and Section 4.4. Furthermore, protocols and challenges for the transport, network, and data-link layers are consecutively explained in Section 4.5 through Section 4.7, respectively.

4.2 Applications/Application Layer Protocols

Sensor nodes can be used for continuous sensing, event detection, event identification, location sensing, and local control of actuators. The concept of microsensing and wireless connection of these nodes promise many new application areas, e.g., military, environment, health, home, commercial, space exploration, chemical processing, and disaster relief, etc. Some of these application areas are described in the next subsection. In addition, Subsection 4.2.2 introduces some application layer protocols used to realize these applications.

4.2.1 Sensor Network Applications

The number of potential applications for sensor networks is huge. Actuators may also be included in the sensor networks, thus making the number of applications that can be developed much higher. In this section, some example applications are given to provide the reader with a better insight about the potentials of sensor networks.

Military applications. Sensor networks can be an integral part of military command, control, communications, computers, intelligence, surveillance, reconnaissance and tracking (C4ISRT) systems. The rapid deployment, self-organization, and fault tolerance characteristics of sensor networks make them a very promising sensing technique for military C4ISRT. Because sensor networks are based on dense deployment of disposable and low-cost sensor nodes, destruction of some nodes by hostile actions does not affect a military operation as much as the destruction of a traditional sensor does. Military applications include: monitoring friendly forces, equipment, and ammunition; battlefield surveillance; reconnaissance of opposing forces and terrain; targeting; battle damage assessment; and nuclear, biological, and chemical attack detection and reconnaissance.

Environmental applications. Some environmental applications of sensor networks include tracking the movements of species, i.e., habitat monitoring; monitoring environmental conditions that affect crops and livestock; irrigation; macroinstruments for large-scale Earth monitoring and planetary exploration; and chemical/biological detection [1, 3, 4, 6, 15, 17, 19, 20, 39, 45].

Commercial Applications: The sensor networks are also applied in many commercial applications, including building virtual keyboards; managing inventory control; monitoring product quality; constructing smart office spaces; and environmental control in office buildings [1, 6, 11, 12, 20, 31, 33, 34, 38, 45].

4.2.2 Application Layer Protocols

Although many application areas for sensor networks are defined and proposed, potential application layer protocols for sensor networks remain largely unexplored. Three possible application layer protocols are introduced in this section: sensor management protocol; task assignment and data advertisement protocol; and sensor query and data dissemination protocol. These protocols may require protocols at other stack layers (explained in the remaining sections of this chapter).

4.2.2.1 Sensor Management Protocol (SMP)

Designing an application layer management protocol has several advantages. Sensor networks have many different application areas; accessing them through networks such as the Internet is the aim in some current projects [31]. An application layer management protocol makes the hardware and software of the lower layers transparent to the sensor network management applications.

System administrators interact with sensor networks by using sensor management protocol (SMP). Unlike many other networks, sensor networks consist of nodes that do not have global ID, and they are usually infrastructureless. Therefore, SMP needs to access the nodes by using attribute-based naming and location-based addressing, which are explained in detail in Section 4.6. SMP is a management protocol that provides software operations needed to perform the following administrative tasks:

- Introducing rules related to data aggregation, attribute-based naming, and clustering to the sensor nodes
- Exchanging data related to location-finding algorithms
- Time synchronization of the sensor nodes
- Moving sensor nodes
- Turning sensor nodes on and off
- Querying the sensor network configuration and the status of nodes, and reconfiguring the sensor network
- Authentication, key distribution, and security in data communications

Descriptions of some of these tasks are given in references 8, 11, 30, 36, and 37.

4.2.2.2 Task Assignment and Data Advertisement Protocol (TADAP)

Another important operation in the sensor networks is interest dissemination. Users send their interest to a sensor node, a subset of the nodes, or the whole network. This interest may be about a certain attribute of the phenomenon or a triggering event. Another approach is the advertisement of available data in which the sensor nodes advertise the available data to the users and the users query the data in which they are interested. An application layer protocol that provides the user software with efficient interfaces for interest dissemination is useful for lower layer operations, such as routing.

4.2.2.3 Sensor Query and Data Dissemination Protocol (SQDDP)

The sensor query and data dissemination protocol (SQDDP) provides user applications with interfaces to issue queries, respond to queries, and collect incoming replies. These queries are generally not issued to particular nodes; instead, attribute-based or location-based naming is preferred. For instance, "the locations of the nodes that sense temperature higher than 70°F" is an attribute-based query. Similarly, "temperatures read by the nodes in Region A" is an example of location-based naming.

Similarly, sensor query and tasking language (SQTL) [37] is proposed as an application that provides even a larger set of services. SQTL supports three types of events, which are defined by keywords *receive*, *every*, and *expire*. The *receive* keyword defines events generated by a sensor node when it receives a message; *every* keyword defines events occurring periodically due to a timer time-out; and *expire* keyword defines events occurring when a timer is expired. If a sensor node receives a message intended for it that contains a script, it then executes the script. Although SQTL is proposed, different types of SQDDP can be developed for various applications. The use of SQDDPs may be unique to each application.

SQDDP provides interfaces to issue queries, responds to queries, and collects incoming replies. Other types of protocols are also essential to sensor network applications: the localization and time synchronization protocols. The localization protocol enables sensor nodes to determine their locations; the time synchronization protocol provides sensor nodes with a common view of time throughout the sensor network. Because many communication protocols require knowledge of location and time, it is important to describe the localization and time synchronization techniques in detail in the following sections before transport, network, and data link protocols are discussed later.

4.3   Localization Protocols

Because sensor nodes may be randomly deployed in any area, they must be aware of their locations in order to provide meaningful data to the users. In addition, location information may be required by the network and data-link layer protocols described in Section 4.6 and Section 4.7, respectively. In order to meet design challenges, a localization protocol must be:

- Robust to node failures
- Less sensitive to measurement noise
- Low error in location estimation
- Flexible in any terrain

Currently, two types of localization techniques address these challenges: (1) beacon based and (2) relative location based. Both techniques may use range and angle estimations for sensor node localization via received signal strength (RSS) [23, 42]; time of arrival (TOA) [13, 41]; time difference of arrival (TDOA); and angle of arrival (AOA).

Current localization methods [27, 36] are based on beacons with position known. The ad hoc localization system (AHLoS) [36] requires few nodes to have known location through GPS or through manual configuration. This allows nodes to discover their location through a two-phase process: ranging and estimation. During the ranging phase, each node estimates the range of its neighbors. The estimation phase then allows neighbors that do not have location to use the range estimated in the ranging phase and the known location of the beacons to estimate their locations.

Also, some methods [5, 6] assume beacon signals at known locations. This assumption may be fine for some applications, but sensor nodes may be deployed in regions in which known location is not possible. As a result, Moses and colleagues are investigating self-localization using sources at unknown locations [27]. Although these authors relax the assumption that beacons require fixed locations, the beacons still need a number of signal sources. These signal sources are deployed in the same region as the sensor nodes and used as references by the neighbor nodes to estimate the unknown locations and orientations from the signal sources.

The work of Moses et al. [27] and Savvides et al. [36] is based on signal sources. Other work [7] estimates locations of the sensor nodes by viewing the location estimation problem as a convex optimization problem because a proximity constraint exists between two nodes, i.e., the range of broadcast. In addition to these localization methods, Patwari and coworkers [28] provide the Cramer–Rao bound of sensor location accuracy based on fixed base stations capable of peer-to-peer time of arrival or received signal strength measurements.

Although beacon-based localization protocols are sufficient for certain sensor network applications, some sensor networks may be deployed in areas unreachable by beacons or GPS; they may be frequently jammed by environmental or manually induced noise. In addition, low-end sensor nodes may exhibit nonlinear device behavior and non-Gaussian measurement noise. To overcome these challenges, the location information is relayed hop by hop from the source to the sink. In order to obtain precise relative location information, the sensor nodes must collaboratively work together to assist each other. Furthermore, energy may be additionally conserved by enabling sensor nodes to track the locations of their neighbor nodes.

This relative localization technique is further explored by the perceptive localization framework (PLF) [43]. In this framework, a node is able to detect and track the location of the neighboring node by using a collaborative estimation technique and a particle filter applied to an array of sensors. To increase the accuracy of the location estimation, the sink may request all the nodes along the path to the sources to increase the number of samples (particles) for particle filtering. This process of local interaction does not require any beacon in place. In addition, a central processing unit is not required in order to determine the locations of the sources.

Whether the beacon- or relative location-based localization protocol is used, the location information is required by the protocols in the transport, network, and data-link layers. Each type of localization protocols offers different capabilities. Future sensor network applications may utilize a combination of localization techniques.

4.4 Time Synchronization Protocols

Instead of time synchronization between the sender and receiver during an application, such as in the Internet, the sensor nodes in the sensor field must maintain a similar time within a certain tolerance throughout the lifetime of the network. Combining with the criteria that sensor nodes must be energy efficient, low cost, and small in a multihop environment as described in Section 4.1, this requirement offers a challenging problem. In addition, the sensor nodes may be left unattended for a long period of time, e.g., in deep space or on an ocean floor. For short-distance multihop broadcast, data processing time and the variation of data processing time may contribute the most in time fluctuations and differences in path delays. Also, the time difference between two sensor nodes is significant over time due to the wandering effect of the local clocks.

Small and low-end sensor nodes may exhibit device behaviors much worse than those of large systems such as personal computers (PCs). Some of the factors influencing time synchronization in large systems also apply to sensor networks [21]:

- *Temperature.* Because sensor nodes are deployed in various places, the temperature variation throughout the day may cause the clock to speed up or slow down. For a typical PC, the clock drifts few parts per million during the day [25]. For low-end sensor nodes, the drifting may be even worse.
- *Phase noise.* Some of the causes of phase noise are due to access fluctuation at the hardware interface, response variation of the operating system to interrupts, and jitter in the network delay. The latter may be due to medium access and queueing delays.
- *Frequency noise.* The frequency noise is due to the instability of the clock crystal. A low-end crystal may experience large frequency fluctuation because the frequency spectrum of the crystal has large sidebands on adjacent frequencies.
- *Asymmetric delay.* Because sensor nodes communicate with each other through the wireless medium, the delay of the path from one node to another may be different from that of the return path. As a result, an asymmetric delay may cause an offset to the clock that cannot be detected by a variance type method [21]. If the asymmetric delay is static, the time offset between any two nodes is also static. The asymmetric delay is bounded by one half the round trip time between the two nodes [21].
- *Clock glitches.* Clock glitches are sudden jumps in time that may be caused by hardware or software anomalies such as frequency and time steps.

Table 4.1 shows three types of timing techniques, each of which must address the challenges mentioned earlier. In addition, the timing techniques must be energy aware because the batteries of the sensor nodes are limited. Also, they must address the mapping between the sensor network time and the Internet time, e.g., universal coordinated time. Next, examples of these types of timing techniques are described, namely, the network time protocol (NTP) [24]; the reference-broadcast synchronization (RBS) [9]; and the time-diffusion synchronization protocol (TDP) [44].

TABLE 4.1   Three Types of Timing Techniques

Type	Description
(1) Relies on fixed time servers to synchronize the network	The nodes are synchronized to time servers that are readily available. These time servers are expected to be robust and highly precise.
(2) Translates time throughout the network	The time is translated hop-by-hop from the source to the sink. In essence, it is a time translation service.
(3) Self-organizes to synchronize the network	The protocol does not depend on specialized time servers. It automatically organizes and determines the master nodes as the temporary time-servers.

In the Internet, the NTP is used to discipline the frequency of each node's oscillator. It may be useful to use NTP to discipline the oscillators of the sensor nodes, but connection to the time servers may not be possible because of frequent sensor node failures. In addition, disciplining all the sensor nodes in the sensor field may be a problem because of interference from the environment and large variation of delay between different parts of the sensor field. The interference can temporarily disjoint the sensor field into multiple smaller fields, causing undisciplined clocks among these smaller fields. The NTP protocol may be considered type 1 of the timing techniques; in addition, it must be refined to address timing challenges in the sensor networks.

The RBS, type 2 of the timing techniques, provides instantaneous time synchronization among a set of receivers within the reference broadcast of the transmitter. The transmitter broadcasts m reference packets. Each of the receivers within the broadcast range records the time of arrival of the reference packets. Afterwards, the receivers communicate with each other to determine the offsets. To provide multihop synchronization, it is proposed to use nodes receiving two or more reference broadcasts from different transmitters as translation nodes. These translation nodes are used to translate the time between different broadcast domains. As shown in Figure 4.1, nodes A, B, and C are the transmitter, receiver, and translation nodes, respectively.

Another emerging timing technique is the TDP, which is used to maintain the time throughout the network within a certain tolerance. The tolerance level can be adjusted based on the purpose of the sensor networks. The TDP automatically self-configures by electing master nodes to synchronize the sensor network. In addition, the election process is sensitive to energy requirement as well as the quality of the

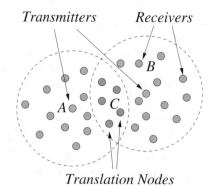

FIGURE 4.1   The RBS.

clocks. The sensor network may be deployed in unattended areas, and the TDP still synchronizes the unattended network to a common time. It is considered type 3 of the timing techniques.

In summary, these timing techniques may be used for different types of applications as discussed in Section 4.2; each has its benefits. A time-sensitive application must choose not only the type of timing techniques but also the type of transport, network, and data-link schemes described in the following sections. This is because different protocols provide different features and services to the time-sensitive application.

4.5   Transport Layer Protocols

The collaborative nature of the sensor network paradigm brings several advantages over traditional sensing, including greater accuracy, larger coverage area, and extraction of localized features. The realization of these potential gains, however, directly depends on efficient, reliable communication between the sensor network entities, i.e., the sensor nodes and the sink. To accomplish this, a reliable transport mechanism is imperative.

In general, the main objectives of the transport layer are (1) to bridge application and network layers by application multiplexing and demultiplexing; (2) to provide data delivery service between the source and the sink with an error control mechanism tailored according to the specific reliability requirement of the application layer; and (3) to regulate the amount of traffic injected into the network via flow and congestion control mechanisms. Nevertheless, the required transport layer functionalities to achieve these objectives in the sensor networks are subject to significant modifications in order to accommodate unique characteristics of the sensor network paradigm. Energy, processing, and hardware limitations of the sensor nodes bring further constraints on the transport layer protocol design. For example, conventional end-to-end, retransmission-based error control mechanisms and window-based, additive-increase, multiplicative-decrease congestion control mechanisms adopted by the vastly used transport control protocol (TCP) may not be feasible for the sensor network domain and thus may lead to waste of scarce resources.

On the other hand, unlike other conventional networking paradigms, the sensor networks are deployed with a specific sensing application objective, such as event detection, event identification, location sensing, and local control of actuators, for a wide range of applications (e.g., military, environment, health, space exploration, and disaster relief). The specific objective of the sensor network also influences the design requirements of the transport layer protocols. For example, the sensor networks deployed for different applications may require different reliability levels as well as different congestion control approaches. Consequently, development of transport layer protocols is a challenge because the limitations of the sensor nodes and the specific application requirements primarily determine design principles of transport layer protocols.

Due to the application-oriented and collaborative nature of the sensor networks, the main data flow takes place in the forward path, where the source nodes transmit their data to the sink. The reverse path, on the other hand, carries the data originated from the sink, such as programming/retasking binaries, queries, and commands to the source nodes. Therefore, different functionalities are required to handle the transport needs of the forward and reverse paths. Transport layer issues pertaining to these distinct cases are investigated separately in the following subsections.

4.5.1   Event-to-Sink Transport

Under the premise that data flows from source to sink are generally loss tolerant, Wan and coworkers questioned the need for a transport layer for data delivery in the sensor networks [32]. Although the need for end-to-end reliability may not exist because of the sheer amount of correlated data flows, an event in the sensor field needs to be tracked with a certain amount of accuracy at the sink. Therefore, unlike traditional communication networks, the sensor network paradigm necessitates an event-to-sink reliability notion at the transport layer [35]. This involves a reliable communication of the event features

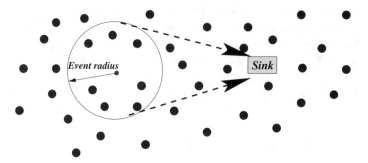

FIGURE 4.2 Typical sensor network topology with event and sink. (The sink is only interested in collective information of sensor nodes within the even radius and not in their individual data.)

to the sink rather than conventional packet-based reliable delivery of the individual sensing reports/packets generated by each sensor node in the field. Figure 4.2 illustrates an event-to-sink reliable transport notion based on collective identification of data flows from the event to the sink.

In order to provide reliable event detection at the sink, possible congestion in the forward path should also be addressed by the transport layer. Once the event is sensed by a number of sensor nodes within the coverage of the phenomenon, i.e., event radius, a significant amount of traffic is triggered by these sensor nodes; this may easily lead to congestion in the forward path. The need for transport layer congestion control to assure reliable event detection at the sink is revealed by the results of Tilak and colleagues [18], who have shown that exceeding network capacity can be detrimental to the observed goodput at the sink. Moreover, although the event-to-sink reliability may be attained even in the presence of packet loss due to network congestion (thanks to the correlated data flows), a suitable congestion control mechanism can also help conserve energy while maintaining desired accuracy levels at the sink.

On the other hand, although the transport layer solutions in conventional wireless networks are relevant, they are simply inapplicable for event-to-sink reliable transport in the sensor networks. These solutions mainly focus on reliable data transport following end-to-end TCP semantics and are proposed to address challenges posed by wireless link errors and mobility [2]. The primary reason for their inapplicability is their notion of end-to-end reliability, which is based on acknowledgments and end-to-end retransmissions. Because of inherent correlation in the data flows generated by the sensor nodes, however, these mechanisms for strict end-to-end reliability are superfluous and drain significant amounts of energy.

In contrast to the transport layer protocols for conventional end-to-end reliability, the event-to-sink reliable transport (ESRT) protocol [35] is based on the event-to-sink reliability notion and provides reliable event detection without any intermediate caching requirements. ESRT is a novel transport solution developed to achieve reliable event detection in the sensor networks with minimum energy expenditure. It includes a congestion control component that serves the dual purpose of achieving reliability and conserving energy. ESRT also does not require individual sensor identification, i.e., an event ID suffices. Importantly, the algorithms of ESRT mainly run on the sink, with minimal functionality required at resource-constrained sensor nodes.

4.5.2   Sink-to-Sensors Transport

Although data flows in the forward path carry correlated sensed/detected event features, the flows in the reverse path mainly contain data transmitted by the sink for an operational or application-specific purpose. This may include operating system binaries; programming/retasking configuration files; and application-specific queries and commands. Dissemination of this type of data mostly requires 100% reliable delivery. Therefore, the event-to-sink reliability approach introduced before would not suffice to address the tighter reliability requirements of flows in the reverse paths.

This strict reliability requirement for the sink-to-sensors transport of operational binaries and application-specific queries and commands involves a certain level of retransmission as well as acknowledgment mechanisms. However, these mechanisms should be incorporated into the transport layer protocols cautiously in order not to compromise scarce sensor network resources totally. In this respect, local retransmissions and negative acknowledgment approaches would be preferable over end-to-end retransmissions and acknowledgments to maintain minimum energy expenditure.

On the other hand, the sink is involved more in the sink-to-sensor data transport on the reverse path, so a sink with plentiful energy and communication resources can broadcast data with its powerful antenna. This helps to reduce the amount of traffic forwarded in the multihop sensor network infrastructure and thus helps sensor nodes conserve energy. Therefore, data flows in the reverse path may experience less congestion compared to the forward path, which is totally based on multihop communication. This calls for less aggressive congestion control mechanisms for the reverse path compared to the forward path in the sensor networks.

Wan and colleagues [32] propose the pump slowly, fetch quickly (PSFQ) mechanism for reliable retasking/reprogramming in the sensor networks. PSFQ is based on slowly injecting packets into the network but performing aggressive hop-by-hop recovery in case of packet loss. The pump operation in PSFQ simply performs controlled flooding and requires each intermediate node to create and maintain a data cache to be used for local loss recovery and in-sequence data delivery. Although this is an important transport layer solution for the sensor networks, PSFQ does not address packet loss due to congestion.

In summary, the transport layer mechanisms that can address the unique challenges posed by the sensor network paradigm are essential to realize the potential gains of the collective effort of sensor nodes. As discussed in the preceding two subsections, promising solutions exist for event-to-sink and sink-to-sensors reliable transports. These solutions and those currently under development, however, need to be exhaustively evaluated under real sensor network deployment scenarios to reveal their shortcomings. Therefore, necessary modifications may be required to provide a complete transport layer solution for the sensor networks.

4.6   Network Layer Protocols

Sensor nodes may be scattered densely in an area to observe a phenomenon. As a result, they may be very close to each other. In such a scenario, multihop communication may be a good choice for sensor networks with strict requirements on power consumption and transmission power levels. As compared to long distance wireless communication, multihop communication may be an effective way to overcome some of the signal propagation and degradation effects. In addition, the sensor nodes consume much less energy when transmitting a message because the distances between sensor nodes are shorter.

As discussed in Section 4.1, ad hoc routing techniques already proposed in the literature [29] do not usually fit requirements of the sensor networks. As a result, the network layer of the sensor networks is usually designed according to the following principles:

- Energy efficiency is always an important consideration.
- Sensor networks are mostly data centric.
- An ideal sensor network has attribute-based addressing and location awareness.
- Data aggregation is useful only when it does not hinder the collaborative effort of the sensor nodes.
- The routing protocol is easily integrated with other networks, e.g., Internet.

These design principles serve as a guideline when designing a routing protocol for sensor networks. Each of them is further explained to emphasize its importance. As described in the preceding section, a transport layer protocol must be energy efficient. This requirement also applies to a routing protocol because the network lifetime depends on the nodes' energy consumption when relaying messages. As a result, energy efficiency plays an important role in various protocol stack layers in addition to the network layer.

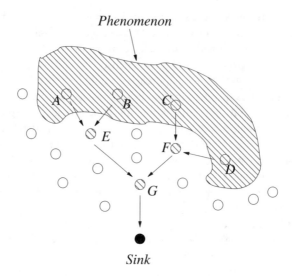

FIGURE 4.3 Data aggregation.

In sensor networks, information or data may be described by using attributes. In order to integrate tightly with the information or data, a routing protocol may be designed according to data-centric techniques. A data-centric routing protocol requires attribute-based naming [8, 10, 26, 37], which is used to carry out queries by using the attributes of the phenomenon. In essence, the users are more interested in the data gathered by the sensor networks in the phenomenon rather than by an individual node. They query the sensor networks by using attributes of the phenomenon that they want to observe. For example, the users may send out a query such as, "find the locations of areas where the temperature is over 70°F."

Furthermore, a data-centric routing protocol should also utilize the design principle of data aggregation — a technique used to solve the implosion and overlap problems in data-centric routing [15]. As shown in Figure 4.3, the sink queries the sensor network to observe the ambient condition of the phenomenon. The sensor network used to gather the information can be perceived as a reverse multicast tree, where the nodes within the area of the phenomenon send the collected data toward the sink. Data coming from multiple sensor nodes are aggregated as if they are about the same attribute of the phenomenon when they reach the same routing node on the way back to the sink. For example, sensor node *E* aggregates the data from sensor nodes *A* and *B* while sensor node *F* aggregates the data from sensor nodes *C* and *D* in Figure 4.3.

Data aggregation can be perceived as a set of automated methods of combining data from many sensor nodes into a set of meaningful information [16]. In this respect, data aggregation is known as data fusion [15]. Also, care must be taken when aggregating data because the specifics of the data, e.g., the locations of reporting sensor nodes, should not be left out. Such specifics may be needed by certain applications.

One of the design principles for the network layer is to allow easy integration with other networks such as the satellite network and the Internet. As shown in Figure 4.4, the sinks are the basis of a communication backbone that serves as a gateway to other networks. The users may query the sensor networks through the Internet or the satellite network, depending on the purpose of the query or the type of application the users are running.

A brief summary of the state of the art in the networking area is shown in Table 4.2. The schemes listed in the table utilize some of the design principles previously discussed. For example, the SMECN [22] creates an energy-efficient subgraph of the sensor networks. It tries to minimize the energy consumption while maintaining connectivity of the nodes in the network. In addition, the directed diffusion protocol [17] is a data-centric dissemination protocol in which the queries and collected data use attribute-based naming schemes.

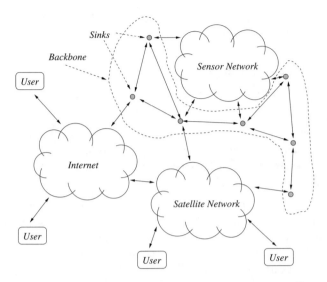

FIGURE 4.4 Internetworking between sensor nodes and user through Internet or satellite network.

TABLE 4.2 Overview of Network Layer Schemes

Network Layer Scheme	Description
SMECN [22]	Creates a sub graph of the sensor network that contains the minimum-energy path.
LEACH [16]	Forms clusters to minimize energy dissipation.
SAR [40]	Creates multiple trees where the root of each tree is one hop neighbor from the sink; select a tree for data to be routed back to the sink according to the energy resources and additive QoS Metric.
Flooding	Broadcasts data to all neighbor nodes regardless if they receive it before or not.
Gossiping [14]	Sends data to one randomly selected neighbor.
SPIN [15]	Sends data to sensor nodes only if they are interested; has three types of messages, i.e., ADV, REQ, and DATA.
Directed Diffusion [17]	Sets up gradients for data to flow from source to sink during interest dissemination.

Because different applications may require different types of network layer protocols, more advanced data-centric routing protocols are needed. In essence, application-specific requirements are part of the driving forces urging for new transport layer protocols, as described in the previous section. In addition, they push for the new data-link schemes described in the following section.

4.7 Data Link Layer Protocols

Although the transport layer mechanisms discussed in Section 4.5 are essential to achieving higher level error and congestion control, it is still imperative to have data-link layer functionalities in the sensor networks. In general, the data link layer is primarily responsible for multiplexing data streams, data frame detection, medium access, and error control; it ensures reliable point-to-point and point-to-multipoint connections in a communication network. Nevertheless, the collaborative and application-oriented nature of the sensor networks and the physical constraints of the sensor nodes, such as energy and processing limitations, determine the way in which these responsibilities are fulfilled. In the following two subsections, data-link layer issues are explored within the discussion of medium access and error control strategies in the sensor networks.

4.7.1   Medium Access Control

The medium access control (MAC) layer protocols in a multihop self-organizing sensor network must achieve two objectives:

- Establish data communication links for creating a basic network infrastructure needed for multihop wireless communication in a densely scattered sensor field.
- Regulate access to shared media so that communication resources are fairly and efficiently shared among the sensor nodes.

Due to the unique resource constraints and application requirements of sensor networks, however, the MAC protocols for the conventional wireless networks are inapplicable to the sensor network paradigm. For example, the primary goal of a MAC protocol in an infrastructure-based cellular system is to provide high QoS and bandwidth efficiency, mainly with dedicated resource assignment strategy. Such an access scheme is impractical for sensor networks because there is no central controlling agent like the base station. Moreover, power efficiency directly influences network lifetime in a sensor network and thus is of prime importance.

Although Bluetooth and the *mobile ad hoc network* (MANET) show similarities to the sensor networks in terms of communication infrastructure, both consist of nodes with portable battery-powered devices that can be replaced by the user. Therefore, unlike in the sensor networks, power consumption is only of secondary importance in these systems. Therefore, none of the existing Bluetooth or MANET MAC protocols can be directly used in the sensor networks because of network lifetime concerns.

It is evident that the MAC protocol for sensor networks must have built-in power conservation, mobility management, and failure recovery strategies. Thus far, *fixed allocation* and *random access* versions of medium access have been proposed [40, 46]. *Demand-based* MAC schemes may be unsuitable for sensor networks due to their large messaging overhead and link setup delay. Furthermore, contention-based channel access is deemed unsuitable because of the requirement to monitor the channel at all times — an energy-draining task.

The applicability of the fundamental MAC schemes in the sensor networks is discussed along with some proposed MAC solutions using that access method as follows:

- *TDMA-based medium access.* Time-division multiple-access (TDMA) access schemes inherently conserve more energy compared to contention-based schemes because the duty cycle of the radio is reduced and no contention-introduced overhead and collisions are present. Pottie and Kaiser [31] have reasoned that a MAC scheme for energy-constrained sensor networks should include a variant of TDMA because radios must be turned off during idling for precious power savings. The self-organizing medium access control for sensor networks (SMACS) [40] is such a time slot-based scheme; each sensor node maintains a TDMA-like super frame in which the node schedules different time slots to communicate with its known neighbors. SMACS achieves power conservation by using a random wake-up schedule during the connection phase and by turning the radio off during idle time slots. However, although a TDMA-based access scheme minimizes the transmit on time, it is not always preferred because of associated time synchronization costs.
- *Hybrid TDMA/FDMA-based medium access.* A pure TDMA-based access scheme dedicates the entire channel to a single sensor node; however, a pure frequency-division multiple access (FDMA) scheme allocates minimum signal bandwidth per node. Such contrast brings the trade-off between the access capacity and the energy consumption. An analytical formula is derived in Shih et al. [38] to find the optimum number of channels, which gives the minimum system power consumption. This determines the hybrid TDMA/FDMA scheme to be used. The optimum number of

channels depends on the ratio of the power consumption of the transmitter to that of the receiver. If the transmitter consumes more power, a TDMA scheme is favored, while the scheme leans toward FDMA when the receiver consumes greater power [38].

- *CSMA-based medium access.* Based on carrier sensing and backoff mechanism, traditional carrier-sense multiple access (CSMA)-based schemes are inappropriate because they make the fundamental assumption of stochastically distributed traffic and tend to support independent point-to-point flows. On the other hand, the MAC protocol for sensor networks must be able to support variable, but highly correlated and dominantly periodic traffic. Any CSMA-based medium access scheme has two important components: the listening mechanism and the backoff scheme. Woo and Culler [46] present a CSMA-based MAC scheme for sensor networks and observe from the simulations that the constant listen periods are energy efficient and the introduction of random delay provides robustness against repeated collisions.

4.7.2 Error Control

In addition to medium access control, error control of the transmitted data in the sensor networks is another extremely important function of the data-link layer. Error control is critical, especially in some sensor network applications such as mobile tracking and machine monitoring. In general, the error control mechanisms in communication networks can be categorized into two main approaches: forward error correction (FEC) and automatic repeat request (ARQ).

ARQ-based error control mainly depends on retransmission for the recovery of lost data packets/frames. It is clear that this ARQ-based error control mechanism incurs significant additional retransmission cost and overhead. Although ARQ-based error control schemes are utilized at the data-link layer for the other wireless networks, the usefulness of ARQ in sensor network applications is limited due to the scarcity of the energy and processing resources of the sensor nodes. On the other hand, FEC schemes have inherent decoding complexity that require relatively considerable processing resources in the sensor nodes. In this respect, simple error control codes with low-complexity encoding and decoding might present the best solutions for error control in the sensor networks.

On the other hand, for the design of efficient FEC schemes, it is important to have good knowledge of channel characteristics and implementation techniques. Channel bit error rate (BER) is a good indicator of link reliability. In fact, a good choice of the error correcting code can result in several orders of magnitude reduction in BER and an overall gain. The coding gain is generally expressed in terms of the additional transmit power needed to obtain the same BER without coding.

Therefore, the link reliability can be achieved by increasing the output transmit power or the use of suitable FEC scheme. Due to energy constraints of the sensor nodes, increasing the transmit power is not a feasible option. Therefore, using FEC is still the most efficient solution, given the constraints of the sensor nodes. Although the FEC can achieve significant reduction in the BER for any given value of the transmit power, the additional processing power consumed during encoding and decoding must be considered when designing an FEC scheme. If this additional power is greater than the coding gain, the whole process is not energy efficient and thus the system is better without coding. On the other hand, the FEC is a valuable asset to the sensor networks if the additional processing power is less than the transmission power savings. Thus, the trade-off between this additional processing power and the associated coding gain should be optimized in order to have powerful, energy-efficient, and low-complexity FEC schemes for error control in the sensor networks.

As researchers continue to investigate new FEC schemes for sensor networks, designers must bear in mind that the new schemes may be application specific. The data-link layer remains a challenging area in which to work because sensor nodes are inherently low end. Combining the low-end characteristic of the sensor nodes with harsh deployed terrains calls for new medium-access as well as error-control schemes.

4.8   Conclusion

An overview of the communication protocols for sensor networks is given in this chapter. Challenges and design guidelines for localization, time synchronization, application layer, transport layer, network layer, and data-link layer protocols are explored. As technology advances in the sensor network area, sensor network technologies may become an integral part of our lives.

Acknowledgment

The authors thank Dr. Ian F. Akyildiz for his encouragement and support.

References

1. Agre, J. and Clare, L., An integrated architecture for cooperative sensing networks, *IEEE Computer Mag.*, 106–108, May 2000.
2. Balakrishnan, H. et al., A comparison of mechanisms for improving TCP performance over wireless links, *IEEE/ACM Trans. Networking*, 5(6), 756–769, December 1997.
3. Bhardwaj, M., Garnett, T., and Chandrakasan, A.P., Upper bounds on the lifetime of sensor networks, *IEEE Int. Conf. Commun. '01*, Helsinki, Finland, June 2001.
4. Bonnet, P., Gehrke J., and Seshadri, P., Querying the physical world, *IEEE Personal Commun.*, 10–15, October 2000.
5. Bulusu, N., Heidemann, J., and Estrin, D., GPS-less low-cost outdoor localization for very small devices, *IEEE Personal Commun.*, 7, 28–34, October 2000.
6. Bulusu, N. et al., Scalable coordination for wireless sensor networks: self-configuring localization systems, *Int. Symp. Commun. Theory Applications (ISCTA 2001)*, Ambleside, U.K., July 2001.
7. Doherty, L., Pister, K.S.J., and Ghaoui, L.E., Convex position estimation in wireless sensor networks, *INFOCOM'01*, Anchorage, AK, April 2001.
8. Elson, J. and Estrin, D., Random, ephemeral transaction identifiers in dynamic sensor networks, *Proc. 21st Int. Conf. Distributed Computing Syst.*, 459–468, Phoenix, AZ, April 2001.
9. Elson, J., Girod, L., and Estrin, D., Fine-grained network time synchronization using reference broadcasts, *Proc. 5th Symp. Operating Syst. Design Implementation (OSDI 2002)*, Boston, MA, December 2002.
10. Estrin, D. et al., Instrumenting the world with wireless sensor networks, *Int. Conf. Acoustics, Speech, Signal Process. (ICASSP 2001)*, Salt Lake City, UT, May 2001.
11. Estrin, D. et al., Next century challenges: scalable coordination in sensor networks, *ACM Mobicom '99*, ,263–270, Seattle, WA, August 1999.
12. Estrin, D., Govindan R., and Heidemann J., Embedding the Internet, *Commun. ACM*, 43, 38–41, May 2000.
13. Fischer, S. et al., System performance evaluation of mobile positioning methods, *Proc. IEEE Vehicular Technol. Conf.*, Houston, TX, May 1999.
14. Hedetniemi, S., Hedetniemi, S., and Liestman, A., A survey of gossiping and broadcasting in communication networks, *Networks*, 18(4), 319–349, 1988.
15. Heinzelman, W.R., Kulik, J., and Balakrishnan, H., Adaptive protocols for information dissemination in wireless sensor networks, *ACM Mobicom '99*, 174–185, Seattle, WA, August 1999.
16. Heinzelman, W.R., Chandrakasan, A., and Balakrishnan, H., Energy-efficient communication protocol for wireless microsensor networks, *IEEE Proc. Hawaii Int. Conf. Syst. Sci.*, 1–10, Maui, HI, January 2000.
17. Intanagonwiwat, C., Govindan, R., and Estrin, D., Directed diffusion: a scalable and robust communication paradigm for sensor networks, *ACM Mobicom '00*, 56–67, Boston, MA, August 2000.
18. Tilak, S., Abu–Ghazaleh, N.B., and Heinzelman, W., Infrastructure trade-offs for sensor networks, *Proc. WSNA 2002*, Atlanta, GA, September 2002.

19. Jaikaeo, C., Srisathapornphat, C., and Shen, C., Diagnosis of sensor networks, *IEEE Int. Conf. Commun. '01*, Helsinki, Finland, June 2001.

20. Kahn, J.M., Katz, R.H., and Pister, K.S.J., Next century challenges: mobile networking for Smart Dust, *ACM Mobicom '99*, 271–278, Seattle, WA, August 1999.

21. Levine, J., Time synchronization over the Internet using an adaptive frequency-locked loop, *IEEE Trans. Ultrasonics, Ferroelectrics, Frequency Control*, 46(4), 888–896, July 1999.

22. Li, L. and Halpern, J.Y., Minimum-energy mobile wireless networks revisited, *IEEE Int. Conf. Commun. ICC '01*, Helsinki, Finland, June 2001.

23. Mark, B. and Zaidi, Z., Robust mobility tracking for cellular networks, *Proc. IEEE Int. Commun. Conf.*, New York, 2002.

24. Mills, D.L., Internet time synchronization: the network time protocol, *Global States and Time in Distributed Systems*, IEEE Computer Society Press, 1994.

25. Mills, D.L., Adaptive hybrid clock discipline algorithm for the network time protocol, *IEEE/ACM Trans. Networking*, 6(5), 505–514, October 1998.

26. Mirkovic, J. et al., A self-organizing approach to data forwarding in large-scale sensor networks, *IEEE Int. Conf. Commun. ICC '01*, Helsinki, Finland, June 2001.

27. Moses, R., Krishnamurthy, D., and Patterson, R., A self-localization method for wireless sensor networks, *Eurasip J. Appl. Signal Process.*, 4, 348–358, 2003.

28. Patwari, N. et al., Relative location estimation in wireless sensor networks, *IEEE Trans. Signal Process.*, August 2003.

29. Perkins, C., *Ad Hoc Networks*, Addison–Wesley, Reading, MA, 2000.

30. Perrig, A. et al., SPINS: security protocols for sensor networks, *Proc. ACM MobiCom '01*, 189–199, Rome, July 2001.

31. Pottie, G.J. and Kaiser, W.J., Wireless integrated network sensors, *Commun. ACM*, 43(5), 551–558, May 2000.

32. Wan, C.Y., Campbell, A.T., and Krishnamurthy, L., PSFQ: a reliable transport protocol for wireless sensor networks, *Proc. WSNA 2002*, Atlanta, GA, September 2002.

33. Rabaey, J. et al., PicoRadio: ad hoc wireless networking of ubiquitous low-energy sensor/monitor nodes, *Proc. IEEE Computer Soc. Annu. Workshop VLSI (WVLSI '00)*, 9–12, Orlando, FL, April 2000.

34. Rabaey, J.M. et al., PicoRadio supports ad hoc ultra-low power wireless networking, *IEEE Computer Mag.*, 33, 42–48, July 2000.

35. Sankarasubramaniam, Y., Akan, O.B., and Akyildiz, I.F., ESRT: event-to-sink reliable transport for wireless sensor networks, *Proc. ACM MOBIHOC 2003*, 177–188, Annapolis, MD, June 2003.

36. Savvides, A., Han, C., and Srivastava, M., Dynamic fine-grained localization in ad hoc networks of sensors, *Proc. ACM MobiCom '01*, 166–179, Rome, July 2001.

37. Shen, C., Srisathapornphat, C., and Jaikaeo, C., Sensor information networking architecture and applications, *IEEE Personal Commun.*, 52–59, August 2001.

38. Shih, E. et al., Physical layer driven protocol and algorithm design for energy-efficient wireless sensor networks, *ACM Mobicom '01*, 272–286, Rome, July 2001.

39. Slijepcevic, S. and Potkonjak, M., Power efficient organization of wireless sensor networks, *IEEE Int. Conf. Commun. '01*, Helsinki, Finland, June 2001.

40. Sohrabi, K. et al., Protocols for self-organization of a wireless sensor network, *IEEE Personal Commun.*, 16–27, October 2000.

41. Spirito, M.A. and Mattioli, A.G., On the hyperbolic positioning of GSM mobile stations, *Proc. Int. Symp. Signals, Syst. Electron.*, September 1998.

42. Spirito, M.A., Further results on GSM mobile station location, *IEEE Electron. Lett.*, 35(22), 1999.

43. Su, W. and Akyildiz, I.F., Perceptive localization framework for sensor networks, *Georgia Tech Technical Report*, 2003.

44. Su, W. and Akyildiz, I.F., Time-diffusion synchronization protocol for sensor networks, *Georgia Tech Technical Report*, 2003.

45. Warneke, B., Liebowitz, B., and Pister, K.S.J., Smart Dust: communicating with a cubic-millimeter computer, *IEEE Computer Mag.*, 2–9, January 2001.
46. Woo, A. and Culler, D., A transmission control scheme for media access in sensor networks, *ACM Mobicom '01*, 221–235, Rome, July 2001.

5

A Comparative Study of Energy-Efficient (E²) Protocols for Wireless Sensor Networks

Quanhong Wang
Queen's University

Hossam Hassanein
Queen's University

5.1 Introduction

A typical wireless sensor network (WSN) may contain hundreds to thousands of microsensor nodes, which are connected by a wireless medium. These sensor nodes are capable of capturing various physical properties, such as temperature, humidity, or pressure, and mapping the physical characteristics of the environment to quantitative measurements. Rapid progress in microelectromechanical system (MEMS) and radio frequency (RF) design, as well as advances in communication protocols and algorithms, have made WSNs more intelligent and led them to ubiquitous deployment.

WSNs exhibit revolutionary approaches to providing reliable, time-critical, and constant environment sensing, event detecting and reporting, target localization, and tracking. Due to their ease of deployment, reliability, scalability, flexibility, and self-organization, WSNs can be deployed in almost any environment, especially those in which conventional wired sensor systems are impossible, unavailable, or inaccessible, such as in inhospitable terrain, dangerous battlefields, outer space, or deep oceans. Therefore, the existing and potential applications of WSNs span a wide spectrum in various domains such as [2, 10, 16, 21–23, 27, 39, 53, 68, 71, 76, 86, 98, 103]:

- Control, communications, computing, intelligence, surveillance, reconnaissance and targeting (C⁴ISRT) for military purposes
- Environmental detection and monitoring

- Disaster prevention and relief
- Medical care
- Home automation
- Scientific exploration
- Interactive surrounding

From a networking architecture perspective, WSNs can be classified as belonging to the family of wireless ad hoc networks, which are collections of wireless, possibly mobile, nodes that are self-configurable to form a network without the aid of any fixed infrastructure. Nodes in the system autonomously handle the necessary control and networking tasks in a distributed manner. The ad hoc architecture overcomes the difficulties raised by the predetermined infrastructure settings of other wireless networks, so a WSN can be randomly and rapidly deployed and reconfigured and easily tailored to specific applications as well. Moreover, ad hoc architecture is highly robust to single node failure and can provide a high level of fault tolerance due to node redundancy and the distributed nature. Furthermore, energy efficiency can be achieved through multihop routing communication. Bandwidth reuse can also benefit from dividing the single long-range hop to multiple short hops; each hop has a considerably short distance [25].

However, because of their unique application requirements, WSNs differ greatly from conventional wireless ad hoc networks [25, 47, 81]. For instance, a WSN usually has a considerably larger number of sensor nodes (hundred to thousands or even more), which is several orders of magnitude greater than in a conventional ad hoc network. The heavy density of nodes leads to high redundancy of data among neighboring nodes. Moreover, a WSN often encounters severe resource constraints, such as power supply, memory, computation speed, etc. Similarly, due to application diversity, the design of a WSN is normally application specific, i.e., it is difficult to devise a unified WSN architecture or deployment strategy to meet the requirements of various applications.

Furthermore, the active duty cycle of a sensor node is fairly low (possibly as low as 1%) and end users generally focus on the collective information, so the data flow is usually unidirectional, i.e., from the sensor nodes to a common processing center. As a result, many existing architectures and protocols for other wireless networks are not suitable for WSNs, and new performance metrics (e.g., system lifetime), in addition to throughput and delay characteristics, should be considered in WSN design. Therefore, novel approaches supporting resource efficiency, scalability, and reliability should be developed to satisfy the specific requirements of WSNs, and numerous research issues remain to be explored.

Section 5.2 of this chapter covers the motivation and directions of energy-efficient protocols with a discussion of QoS metrics and analysis of energy-consuming sources in WSNs. In Section 5.3, the concept of a cross-layer protocol stack dedicated for WSNs is introduced. Section 5.4 classifies and compares various MAC layer protocols targeting energy-efficient and reliable packet transmission. In Section 5.5, a comparative study is carried out on a number of energy-efficient network layer protocols. Section 5.6 concludes the chapter.

5.2 Motivations and Directions

A typical sensor node is compact, tiny, and inexpensive, but it integrates the functionalities of sensing, data processing and computation, and communication. It is normally operated by an attached power supply that is usually a nonrechargeable or nonreplaceable battery [1, 23, 60].

5.2.1 Necessity of Resource Efficiency

The limited physical size of sensor nodes has the inherent problem of severe resource limitation. Therefore, in WSNs, resource efficiency is extremely critical despite its complexity. Above all, energy-efficient protocols are in high demand in order to extend the lifetime of the system. Because a WSN often operates in a human-unattended manner, the power supply (which is usually an attached battery) cannot be replenished in most cases. In addition, efforts should be made to increase efficiency for the utilization of other resources. For example, using algorithms with low complexity will reduce computation time and thus save power. It also decreases the latency of data delivery. Bandwidth-efficient architectures and protocols can accelerate data delivery as well.

It should be noted that it is difficult to issue a unique definition of the system lifetime for all application scenarios. On one hand, a system lifetime can be measured by the time when the first node exhausts its energy, or a system can be declared dead when a certain fraction of nodes die, or even when all nodes die. Using one definition or another depends on the particular application. On the other hand, the system lifetime can also be measured by application-specific parameters, such as the time until the system can no longer provide acceptable results.

5.2.2   QoS with Energy Efficiency Constraints

Quality of service in WSNs can be evaluated by the following metrics [1, 11, 23, 25, 30, 32, 46, 60, 67, 85, 95, 105]:

- *Energy efficiency.* This determines the system lifetime and is a crucial issue in WSNs. It is clear that for the same sensing task, the higher the energy efficiency is, the longer the system will survive.
- *Accuracy.* This reflects the basic value of gathered information because the amount of received data determines the level of accuracy. In general, the more data received, the higher the accuracy should be.
- *Latency.* In most cases, information collected from the monitoring environment is time critical, so it should be delivered in a timely fashion.
- *Security.* Because many WSNs are used for military or surveillance purposes, denial of service attacks against these networks may cause severe damage to their operation. Therefore, data privacy and safe communications are of utmost importance.
- *Fault tolerance.* Although the wireless communication channel is usually noisy, prone to errors, and time varying, data must be delivered reliably. In such cases, data verification and correction on each layer of the network are critical to provide accurate results. Moreover, some sensor nodes may fail due to energy exhaustion or physical obstacles in the environment, so sensor nodes are expected to perform self-testing, self-calibration, self-repair, and self-recovery procedures.
- *Scalability and flexibility.* The system should be scalable and flexible to the enlargement of the network scale. The approaches to scalability and flexibility include clustering, multihop delivery, localization of computation, and data processing.

However, it is impossible to achieve all of these objectives at the same time because some of them conflict with each other. In terms of resource consumption, it is necessary to make a trade-off between energy efficiency and other metrics. Essentially, any of the preceding objectives except the first one is resource hungry. For example, high accuracy requires the delivery of large amounts of data, which leads to more power and bandwidth consumption. Similarly, approaches aiming for timely delivery, security, and reliability are bound to cost extra energy. Local computation is helpful to eliminate the amount of data transmitted, but complex and memory costly computation may cause long latency, and increased power consumption.

5.2.3   Energy Consumption in WSNs

As a microelectronic device, the main task of a sensor node is to detect phenomena, carry out data processing timely and locally, and transmit or receive data. A typical sensor node is generally composed of four components [4, 37, 45, 58, 64, 65, 68, 83, 94, 106]: a power supply unit; a sensing unit; a computing/processing unit; and a communicating unit. The sensing node is powered by a limited battery, which is impossible to replace or recharge in most application scenarios. Except for the power unit, all other components will consume energy when fulfilling their tasks. Extensive study and analysis of energy consumption in WSNs are available [69, 80, 83, 87].

5.2.3.1 Sensing Energy

The sensing unit in a sensor node includes the embedded sensor and/or actuator and the analog–digital converter. It is responsible for capturing the physical characteristics of the sensed environment and converts its measurements to digital signals, which can be processed by a computing/processing unit.

Energy consumed for sensing includes: (1) physical signal sampling and conversion to electrical signal; (2) signal conditioning; and (3) analog to digital conversion. It varies with the nature of hardware as well as applications. For example, interval sensing consumes less energy than continuous monitoring; therefore, in addition to designing low-power hardware, interval sensing can be used as a power-saving approach to reduce unnecessary sensing by turning the nodes off in the inactive duty cycles. However, there is an added overhead whenever transiting from an inactive state to the active state. This leads to undesirable latency as well as extra energy consumption. However, sensing energy represents only a small percentage of the total power consumption in a WSN. The majority of the consumed power is in computing and communication, as discussed next.

5.2.3.2 Computing Energy

The computing/processing unit is a microcontroller unit (MCU) or microprocessor with memory. It carries out data processing and provides intelligence to the sensor node. A real-time micro-operating system running in the computing unit controls and operates the sensing, computing, and communication units through microdevice drivers and decides which parts to turn off and on [11, 36, 38, 58, 69, 79, 82, 87].

Total computing energy consists of two parts: switching energy and leakage energy. The switching energy is determined by supply voltage and the total capacitance switched by executing software. The pattern of draining the energy from the battery affects the total computing energy expense. For example, a scheme of energy saving on computation is dynamic voltage scaling (DVS) [12, 29, 63, 69], which can adaptively adjust operating voltage and frequency to meet the dynamically changing workload without degrading performance. The leakage energy refers to the energy consumption while no computation is carried out. Some researchers have reported that it can reach 50% of the total computing energy. Therefore, it is critical to minimize leakage energy [8, 12, 29, 63, 69].

The concept of system partitioning [11, 59, 93] can also be used to reduce computing energy in sensor nodes. Two practical approaches include removing the intensive computation to a remote processing center that is not energy constrained, or spreading some of the complex computation among more sensors instead of overloading several centralized processing elements.

Energy expenditure for computing is much less compared to that for data communication. Experiments show that the ratio of communicating 1 bit over the wireless medium to that of processing the same bit could be in the range of 1000 and 10,000 [102]. Therefore, trading complex computation/data processing for reducing communication amount is effective in minimizing energy consumption in a multihop sensor network.

5.2.3.3 Communicating Energy

The communicating unit in a sensing node mainly consists of a short-range RF circuit that performs data transmission and reception. The communicating energy is the major contributor to the total energy expenditure and is determined by the total amount of communication and the transmission distance. As reported in Pottie and Kaiser [65], processing data locally to reduce the traffic amount may achieve significant energy savings. Moreover, according to Rappoport [70], signal propagation follows as exponential law to the transmitting distance (usually with exponent 2 to 4 depending on the transmission environment). It is not hard to show that the power consumption due to signal transmission can be saved in orders of magnitude by using multihop routing with a short distance of each hop instead of single-hop routing with a long-distance range for the same destination.

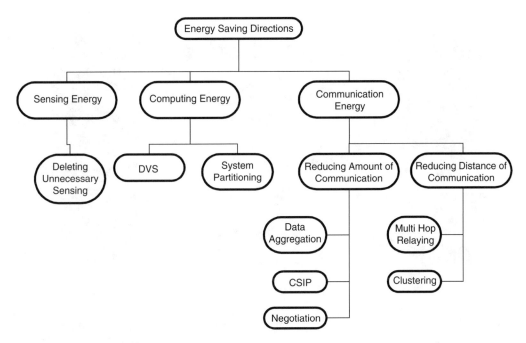

FIGURE 5.1 Energy-conserving directions in WSNs.

Therefore, minimizing the amount of data communicated among sensors and reducing the long transmitting distance into a number of short ones are key elements to optimizing the communicating energy; numerous efforts have focused on these objectives. Several approaches have been devised in order to reduce data communication. For instance,

- Data aggregation has been applied to eliminate redundancy in neighboring nodes [42, 44, 107].
- Collaborative signal and information processing (CSIP) has been used to fulfill local data processing [20, 23, 49, 66, 67, 102, 108].
- Negotiation-based protocols have been introduced to reduce unnecessary replicated data [31, 48].

Similarly, in order to decrease signal transmission distance, multihop communication and clustering-based hierarchies have been proposed to forward data in the network [14, 34, 54, 56, 96].

Figure 5.1 summarizes energy-conserving directions with respect to optimizing sensing, computing, and communication energy consumption. Such approaches exhibit a high degree of dependency on one another. For example, eliminating unnecessary sensing could reduce data communication; in turn, communication energy consumption is reduced. However, this requires more sophisticated control schemes, which are supported by higher complexity computation, and may result in higher energy use for computation. Therefore, trade-offs should be made and some specific direction may take greater importance based on the nature of the application scenario. The remainder of this chapter introduces a number of energy-efficient protocols, which concentrate in one or more of the three directions.

5.3   Cross-Layer Communication Protocol Stack for WSNs

The conventional wireless ad hoc network protocol design is mainly based on a layered stack in which each layer is designed and operated in isolation. The interfaces between layers are static and independent of the individual network constraints and applications. By using this paradigm, network design can be greatly simplified. However, this approach's lack of flexibility and optimality may result in poor performance in large-scale WSNs in which resource limitation is severe, but timely delivery is required [27].

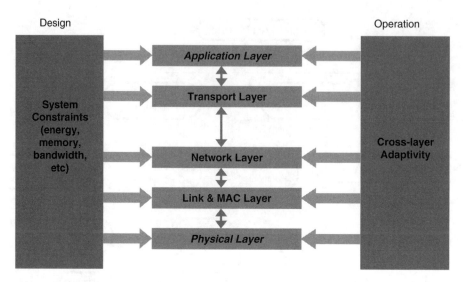

FIGURE 5.2 Cross-layer protocol stack in WSNs.

Therefore, an active theme — cross-layer design — has been recently proposed; this supports optimization and adaptability across multiple layers [27, 33, 73]. A possible cross-layer architecture is depicted in Figure 5.2.

In the concept of cross-layer design, each layer is not developed in isolation, but in an integrated and hierarchical framework. Therefore, the strict border between different layers is loosened. Some control messages as well as information concerning a layer's status will be exchanged among different layers so that the system can take advantage of the interdependencies between them. For example, the link layer can adjust rate, power, and coding to satisfy application requirements based on current channel and network conditions; MAC layer can be adaptive to underlying link and interference conditions, delay constraints, and bit priorities; Routing protocols can be developed according to up-to-date link, network, and traffic conditions; The application layer can adopt the concept of soft QoS, which is adaptive to the underlying network conditions to deliver the highest possible application quality [27].

In practice, cross-layer design may be exercised in some, rather than all, layers in the protocol stack. Discussion will focus on protocols with cross-layer design on network and MAC layers. However, many open problems exist concerning how to understand and implement this concept, what kind of information should be exchanged between layers, and what kinds of internal and external constraints should be taken into consideration.

5.4 Energy-Efficient MAC Protocols

The functions of the data link layer include framing and link access, reliable delivery, flow control, error detection, and retransmission. Because nodes share a common wireless medium for communication, MAC sublayer protocols are critical to providing coordination among nodes. These protocols attempt to provide reliable communication and achieve high throughput with bounded latency, while at the same time minimizing collisions and energy dissipation [50, 89, 90]. The following discussion covers sources influencing energy consumption at the MAC layer, which may lead to directions to improve energy efficiency. Different kinds of energy-efficient MAC layer approaches will also be discussed and some comparisons made in terms of energy efficiency.

5.4.1 Sources of Energy Consumption at the MAC layer

From the point view of energy dissipation, four major sources of energy waste are caused by MAC layer problems [100]:

- *Retransmission due to collision or congestion.* In WSNs, all nodes are capable of transmitting data through the same broadcast channel. As a tiny communication device, each sensor node may have only one receiving antenna; therefore, if two or more transmissions from multiple sources arrive at the same time, a collision will happen, and none of transmitted packets can be received correctly. To ensure reliable transmission, after source nodes detect data collision, they must retransmit, which causes extra energy expenditure. On the other hand, because of limited capacity of the wireless channel, data losses take place when traffic is heavy and the network encounters congestion. This case also requires retransmissions.
- *Idle channel sensing.* In order to eliminate or reduce collisions, nodes must sense the channel continuously to obtain scheduling information or wait before sending data until the channel is detected idle. In either case, extra sensing energy is needed. Indeed, in ad hoc networks, idle channel sensing energy is not negligible compared to data receiving and transmitting. According to Chen and colleagues [14], the ratios of E_{idle}:$E_{receiving}$:$E_{transmitting}$ are 1:2:2.5.
- *Overhearing.* When sharing a common wireless medium, the data transmitted by one node can reach all the other nodes within their transmission range. A node then may receive packets not destined for it. This is referred to as overhearing and it also wastes energy.
- *Overhead due to control messages.* A lot of MAC protocols operate by exchanging control messages for signaling, scheduling, and collision avoidance, which will consume extra energy.

Therefore, in order to design an energy-efficient MAC protocol, collisions must be avoided as much as possible. Moreover, energy dissipation due to idle channel sensing, overhearing, and overhead should also be reduced to a minimum. Many approaches have been proposed, but it is difficult to achieve all energy-conserving objectives at the same time.

5.4.2 Classification and Comparison of MAC Protocols

In general, wireless communication has a variety of MAC protocols, which can be classified into distinct groups according to different criteria. Based on whether a central controller is involved in coordination, WSNs' MAC protocols can be categorized as centralized, distributed (decentralized), and hybrid. Actually, hybrid protocols attempt to combine the advantages of centralized and distributed schemes, but can be more complex. Figure 5.3 shows such classification [97].

5.4.2.1 Centralized MAC Protocols

Centralized MAC protocols include polling algorithms and controlled multiplexing (or channel partitioning) algorithms [50, 70]. A centralized controller is needed to coordinate channel access among the different nodes and collision-free operation can be achieved. Thus, energy wasted due to collisions can be eliminated. However, because of the high overhead and long delay, pure polling mechanisms are not suitable in large-scale WSNs. Depending on how bandwidth is assigned, controlled multiplexing mechanisms can be frequency division multiplexing access (FDMA), code division multiplexing (CDMA), or time division multiplexing access (TDMA). This class of protocols is preferable in WSNs [65], not only because it is collision free, but also because nodes can be turned off in unassigned slots, thus saving energy expenditure due to idle sensing and overhearing.

However, drawbacks exist in channel partitioning schemes. When using TDMA, the central controller will consume more energy than other nodes, and scheduling tends to be dynamic, which will lead to a more complex mechanism. Moreover, it also requires clock synchronization among all nodes, which will

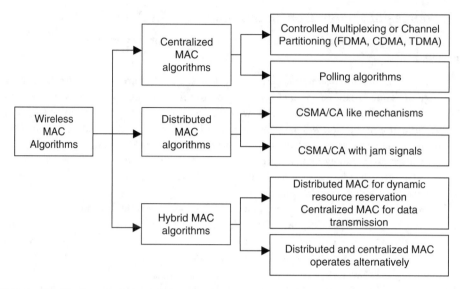

FIGURE 5.3 Classification of wireless MAC protocols.

also dissipate some extra energy. For FDMA, due to the limited bandwidth in the system, it is not realistic to assign a unique frequency for each individual node. Furthermore, bandwidth wastage will occur due to the low duty cycle. Similarly, when using CDMA, although all nodes can transmit at will, some overhead will result because each node must encode its data bits with its uniquely assigned code.

Centralized multiplexing access, therefore, lacks flexibility and scalability to adapt to the variation of WSN applications. Some efforts have been made to improve the performance in terms of energy efficiency. One way is to combine TDMA with other controlled multiplexing, such as self-organizing medium access control for sensor networks (SMACS) [85], which is a combination of TDMA/FDMA MAC protocol. Low-energy adaptive clustering hierarchical (LEACH) [32–34], on the other hand, combines TDMA and CDMA protocols, i.e., it uses TDMA protocol to prevent intracluster* collisions and CDMA to avoid intercluster collisions. Adaptive periodic threshold-sensitive energy-efficient sensor network protocol (APTEEN), described in Manjeshwar and Agrawal [55] and Manjeshwar et al. [56], also uses TDMA with CDMA; however, it adopts a modified TDMA in which the length of time slots assigned to idle nodes and sleeping nodes is different, and all the idle node slots are ordered to precede sleeping nodes. Another alternative is to apply dynamic reservation TDMA (DR-TDMA) [109], which is actually a hybrid approach combining TDMA and carrier sense multiple access (CSMA) mechanisms.

5.4.2.2 Distributed MAC Protocols

Distributed MAC protocols usually provide random multiple access to a wireless medium. Most prevailing MAC protocols in this category adapt carrier sensing and collision avoidance, i.e., based on CSMA/CA. Through carrier sensing, significant transmission collisions can be eliminated by deferring transmission when the channel is detected busy. To further decrease the probability of collision, some collision avoidance measures can be taken, such as a random back-off procedure; a representative example of CSMA/CA based MAC protocol is specified in IEEE 802.11 distributed coordination function (DCF) [41].

However, in some cases, location-dependent carrier sensing results in "hidden" and "exposed" terminal problems, which have a great impact on efficiency. A hidden terminal refers to the node within the range of the intended destination but out of range of the sender, so the hidden terminal cannot be aware of the ongoing transmission. An exposed terminal is the node within the range of the sender but out of range of the destination, so the exposed terminal will be improperly precluded from sending in order to avoid collision. Two types of CSMA/CA-based schemes have been proposed to solve these problems. In

---

*The concept of clusters will be explained later in the chapter.

DCF, the exchange of "request to send–clear to send" (RTS-CTS) control messages reserves the transmission space for subsequent data exchange, thereby eliminating hidden terminal transmission. Deng and Hass [18] and Hass and Deng [104] propose a scheme, called dual busy tone multiple access (DBTMA), that separates control and data channels to relieve the problems raised by hidden and exposed terminals by indicating the transmission or receiving status explicitly.

Other distributed MAC protocols use a jamming signal, such as elimination yield-non-preemptive priority multiple access (EY-NPMA) [3, 24], which is used in the HIPERLAN system (being developed in Europe), and black burst (BB) [84, 110], which is proposed to support prioritized data transmission in ad hoc networks.

Using distributed MAC protocols, nodes operate in a decentralized manner, so it is easy to implement and perform more flexible and scalable control mechanisms, which may fit well with the requirements of WSNs. However, they are not collision-free protocols, and the listen-before-talk scheme calls for all nodes to keep sensing the channel. This results in high energy wastage due to collisions, idle listening, overhearing, and control message overhead. In Ye et al. [100], a novel MAC protocol called sensor-MAC (S-MAC) is proposed and attempts to reduce all four types of energy wastage.

5.4.2.3 Hybrid MAC Protocols

As discussed in previous subsections, conventional centralized and distributed MAC layer protocols cannot provide optimal results in terms of energy efficiency in WSNs. Hybrid MAC protocols attempt to integrate the controllability of centralized protocols with the flexibility of distributed protocols. A number of these protocols are discussed here.

DR-TDMA [109] was originally proposed for wireless ATM networks, but can be extended to WSNs. Figure 5.4 demonstrates the frame structure of the DR-TDMA. Specifically, the fixed-length frame is divided into uplink and downlink time intervals. During the contention phase of the uplink interval, a distributed collision-based MAC scheme — the framed pseudo-Bayesian priority aloha protocol — is used for nodes to transmit temporary reservation requests for next frame to the base station. Uplink data are transmitted in TDMA mode based on the time slots assigned by the base station in the preceding downlink interval. In the downlink interval, a centralized MAC protocol is used to carry out slot assignments, as well as data transmission from base station to nodes. The resource reservation and assignment are adjusted dynamically based on work load. The nodes can turn off during periods outside assigned transmission slots or contention slots. Therefore, energy wastage due to data collision, idle sensing, and overhearing can be reduced. Dynamic slot assignment can also provide flexibility to WSNs.

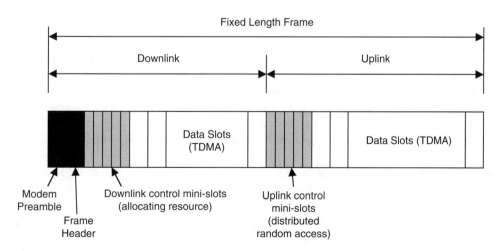

FIGURE 5.4 Frame structure of DR-TDMA.

Two other hybrid TDMA-based protocols are time reservation using adaptive control for energy efficiency (TRACE) [90] and multihop TRACE (MH-TRACE) [91]. Similarly to DR-TDMA, a central controller is in charge of arranging the TDMA transmission schedule according to continuing reservations and new reservation requests. Data are transmitted based on the transmission schedule, which is updated dynamically. The reservation requests are transmitted through a contention-based distributed MAC protocol. Because TRACE and MH-TRACE are dedicated for energy efficiency, these two approaches have a dynamic central controller as opposed to using a fixed-base station as the central controller in DR-TDMA.

A cluster formation scheme is used to manage the nodes. Each cluster head also plays the role of a TDMA scheduler within its cluster. By dynamically choosing cluster heads, balanced energy consumption among the nodes can be achieved. Moreover, these protocols introduce a novel control message — information summarization (IS) — to obtain information on future data transmission within the transmission range of nodes. This way energy wastage due to idle channel sensing and overhearing can be avoided. Due to their characteristics of energy efficiency, flexibility, and self-configuration, these two hybrid MAC protocols look promising for future WSNs. The issues of how to select central controllers dynamically and how to create data transmission schedules are major challenges for such protocols.

5.5   Energy-Efficient Network Layer Protocols

The network layer in WSNs is responsible for data delivery from source to destination via well-selected routes [50, 89]. Due to the unique characteristics of WSNs, many of the network layer protocols designed for conventional ad hoc networks may not fit with the requirements of WSNs. The following principles must be considered in WSN network layer protocols:

- Energy efficiency is always a dominant consideration.
- Routing is often data centric.
- Data aggregation/fusion is desirable, but only useful if it does not affect the collaborative efforts among sensor nodes.
- An ideal sensor network has attribute-based addressing and location awareness.
- Protocols are most likely application specific.

5.5.1   Classification of Network Layer Protocols

To reduce communication's energy consumption, network layer protocols have drawn considerable attention. Many factors influence the design of network layer protocols, and a wide range of schemes have been proposed. Table 5.1 presents a classification of energy-efficient (E^2) network layer protocols. Note that the purpose of such classification is to aid with the study of energy-efficient network layer protocols. Other researchers may opt to use different, and possibly more elaborate, classifications.

In general, the base station, which plays the role of data gathering and processing, may be located far from the sensing field, or can be placed within the network. Sensor nodes can be deployed in the sensing field according to various strategies [19, 92]. This can be predetermined [19, 75, 76] — sensor nodes are placed in preplanned positions, or fixed regular topology or self-regulated [15, 40] — sensor nodes can be spread automatically into the sensing area in sequential steps. Sensor node deployment may also follow a random [32–34, 92] or a biased distribution [94]. The sensor nodes within one system may be identical or heterogeneous in terms of functionality, and resources. The status of the network can be stationary or may change dynamically. Dynamic sensor networks may have mobile sensor nodes, end-users who are collecting the information, or sensed targets [92].

From the data perspective, WSN protocols are designed for global data delivery or local data processing. A detailed classification of the actions on data is shown in Figure 5.5. From the application perspective, the communication schemes can be grouped into three categories [54–56]. The first kind is proactive (also known as source initiated): the sensor nodes keep sensing the environment and continuously report

TABLE 5.1   Classifications of Network Layer Protocols in WSNs

Criteria	Classification		
Position of base station	Far from WSN		
	Within WSN		
Sensor deployment	Predeterministic		
	Self-regulated		
	Random distribution		
	Biased distribution		
Node properties	Homogenous		
	Heterogeneous		
Network dynamics	Static		
	Dynamic	Sensors	
		End-users	
		Targets	
Actions on data	Delivering	Information collection	
		Information dissemination	
		Hybrid	
	Processing	Data Aggregation	
		Collaborative signal and	Event driven
		information Processing	Mobile agent based
		(CSIP)	Relation based
Effective range of the protocols	Globalization (data forwarding)		
	Localization (data processing)		
System architecture	Flat		
	Hierarchical		
Routing approaches	Flooding		
	Unicast		
	Multicast		
Application scenarios	Proactive (source initiated, continuously)		
	Reactive	Phenomenon/event-driven	
		Query-driven (end-user initiated, request-reply on demand)	
	Hybrid		
Energy-efficiency objectives	Minimizing energy consumption in forwarding each individual packet		
	Minimizing in-network total energy consumption		
	Balancing in-network power consumption		

the sensed values. The second kind is reactive: transmission of sensed values is triggered by some specific conditions. They can be driven by a phenomenon or query. The last kind is hybrid, which is a combination of the proactive and reactive methods. In order to realize energy efficiency, different communication protocols have been proposed to fit each scenario.

The architecture of WSNs can be flat or hierarchical. Hierarchical structures can be cluster based, tree based and hierarchical chain based. Routing protocols for efficient data delivery can be flooding based, unicast, or multicast. Furthermore, different kinds of energy efficiency objectives exist, such as minimizing energy dissipation to deliver each individual packet; minimizing in-network total energy dissipation; and balancing in-network energy dissipation.

Based on these classifications, comparisons of some existing network layer protocols that claim to offer energy efficiency will be provided. These protocols are designed for WSNs or were originally designed for wireless ad hoc networks, but can scale to WSNs.

5.5.2   Energy-Efficient Data Delivery Protocols

One of the critical responsibilities of network layer protocols is to provide data delivery between desired source and destination. In WSNs, data delivery protocols should take energy efficiency into consideration. A number of protocols target E^2 data forwarding. These can be classified into distinct groups according

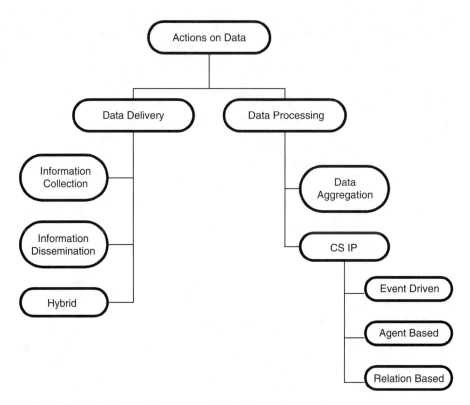

FIGURE 5.5   Classification of network layer protocols based on actions on data.

to several criteria, such as the purpose of flow delivery, application scenario, routing approach, and E^2 objectives. Figure 5.6 demonstrates such classification. Later in this chapter, a case study of data delivery protocols for the purpose of information collection and information dissemination will be provided.

5.5.2.1   Energy-Efficient Information Collection (E²IC) Protocols

Information gathering is one of the essential tasks for all WSNs. A single data-collecting and -processing center is usually assumed, so data are delivered in a unidirectional manner. Therefore, the design of information collection protocols with energy constraints may differ greatly from conventional ad hoc routing protocols.

From the perspective of system architecture, E²IC protocols can be flat or hierarchical, depending on whether the system is divided into space division groups. Figure 5.7 shows this classification of E²IC protocols.

Flat multihop E²IC protocols. The geographical adaptive fidelity (GAF) algorithm [96] and a protocol called SPAN [14] are proposed for wireless ad hoc networks. Due to their scalability, they are also applicable in WSNs. Taking advantage of redundant deployment of sensor nodes and low duty cycle, both protocols designate to rotate switching nodes between active and inactive states without losing the connectivity of the system. In GAF, equivalent nodes are identified based on geographic locations on a virtual grid, so they can substitute each other directly and transparently without affecting the routing topology. Therefore, only one node in a virtual grid needs to be on duty at any time, while all others can go to sleep. In this case, little energy is used, so energy consumption can be reduced.

In SPAN, a limited number of nodes are randomly self-selected as coordinators to construct a backbone in a peer-to-peer fashion within the network for traffic forwarding, while others can make local decisions to transit to a sleep state or keep active. Because a WSN is only sensing its environment and waiting for an interesting event to happen, a new technique described in Schurgers et al. [77, 78] — sparse topology and energy management (STEM) — claims to improve beyond SPAN and GAF in terms of obtaining

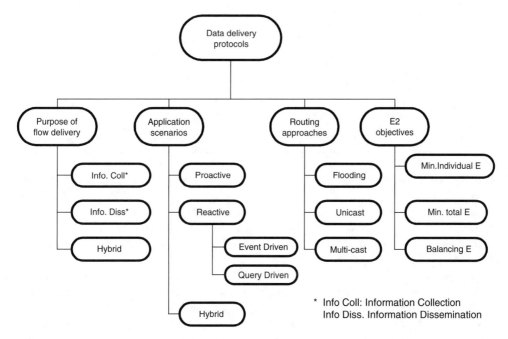

FIGURE 5.6 Classification of data delivery protocols in WSNs.

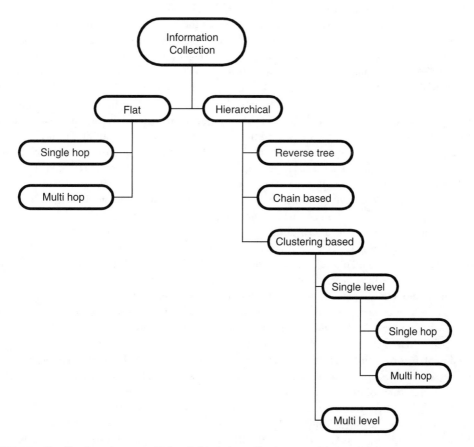

FIGURE 5.7 Classification of energy-efficient information collection protocols in WSNs.

TABLE 5.2   Comparison of Cellular Networks and Clustering-Based WSNs

	Cellular	WSNs
Peer unit	Cell	Cluster
Data processing center (DPC)	Base station	Cluster head
Location of DPC	Fixed	Randomized
Location and form of peer unit	Static	Dynamic
Node positions	Mobile	Static and/or mobile
Interpeer unit communication	Through MSC	Self-configurable

higher energy savings so as to prolong system lifetime by trading off an increased latency to establish a multihop path.

Hierarchical E²IC protocols. Depending on how the hierarchical structure is formed, hierarchical E²IC protocols can be grouped as reserved tree based, chain based, or clustering based. Among these, the clustering-based approach has received increased attention because of its effectiveness, lower complexity, and flexibility.

From the perspective of space division, cluster-based WSNs are similar to cellular networks, although many differences exist between them. For example, in cellular networks, a base station has no resource constraints and is placed at a fixed position, so the cell is static and the nodes there have high mobility. In WSNs, a cluster head (the peer unit of a base station in cellular networks) is generally a sensor node, which has severe resource limitations, and cluster heads are selected dynamically; therefore, clusters are not static within the network, but sensor nodes are often in stationary position. Table 5.2 shows the differences between cellular networks and cluster-based WSNs.

Dividing the entire system into distinct clusters replaces the one-hop long distance transmission by multihop short-distance data forwarding. This would reduce the energy consumed for data communications. Clustering-based E²IC schemes also have the advantages of load balancing, and scalability when the network size grows. Challenges faced by such clustering-based approaches include how to select the cluster heads and how to organize the clusters. The clustering strategy could be single-hop cluster or multihop cluster, based on the distance between the cluster heads and their members. According to the hierarchy of clusters, the clustering strategies can also be grouped into single-level or multilevel clustering.

Various clustering approaches for wireless ad hoc and/or sensor networks have been proposed in the literature. In Heinzelman et al. [32, 33] and Heinzelman [34], the authors propose a distributed LEACH. Initially, each node self-selects itself as a cluster head with a predetermined probability; the cluster head then advertises its decision to other nodes that would make the decision to join a specific cluster that requires minimum communication energy. In order to ensure balanced energy dissipation among all nodes, LEACH invokes the rotation of cluster heads by periodically calling the self-selection and cluster formation procedure. LEACH is a well developed clustering-based protocol dedicated for continuous E²IC in WSNs. However, it is used for proactive application scenarios and does not take the energy consumption for idle sensing of the channel into account; the formation of clusters is not energy aware. Therefore, some efforts have been made to improve its performance further.

Manjeshwar and Agrawal [54] propose the threshold-sensitive energy-efficient sensor network (TEEN) protocol. TEEN adopts the cluster formation method of LEACH, but uses thresholds to achieve enhanced control on sensor nodes. This scheme can also save energy consumption due to idle sensing. It is suitable for time-critical data delivery in reactive application scenarios. Adaptive periodic TEEN (APTEEN) is proposed in Manjeshwar and Agrawal [55] and Manjeshwar et al. [56] to fit in the requirements of hybrid application scenarios using enhanced query management and a modified TDMA MAC protocol. In TEEN and APTEEN, the concept of multilevel clustering is used.

A chain-based protocol called power-efficient gathering in sensor information systems (PEGASIS) is presented in Lindsey and Raghavendra [51] and Lindsey et al. [52]. Instead of sending data packets directly to cluster heads, as is done in the LEACH protocol, each node forwards its packets to the

TABLE 5.3   Comparison of Several Hierarchical Information Collection Protocols

	LEACH	LEACH-C	TEEN	APTEEN	PEGASIS	PA-VBS
Centralized/distributed	D	C	D	C	D	D
Cluster head selection	Self-selection	Nominated by controller	Same as LEACH	Same as LEACH-C	N/A = Not Applicable	Self-selection
Energy awareness in cluster head selection	No	Yes	No	Yes	N/A	Yes
Scalability to large heterogeneous networks	No	Yes	No	Yes	No	Yes
Location awareness	Yes	Yes	Yes	Yes	Yes	Yes
Balanced energy dissemination	Yes	Yes	Yes	Yes	Yes	Yes
Adoption of data aggregation	Yes	Yes	Yes	Yes	Yes	No
Cross-layer design	Yes	Yes	Yes	Yes	Yes	Yes

destination through its closest neighbors. Utilizing the feature of randomized creation and rotation of cluster heads as proposed in LEACH, as well as the advantages of multihop clustering algorithms, Bandyopadhyay and Coyle [7] introduced a new energy-efficient, single-level, multihop clustering algorithm.

These authors [7] also provide a formulation for finding the optimal parameter values to minimize energy consumption, as well as a novel energy-efficient hierarchical clustering algorithm with a total of h levels, i.e., some of the cluster heads in level $k - 1$ select themselves as kth level cluster heads, and the remaining level $k - 1$ cluster heads are cluster members in level k. Based on the results of Foss and Zuyev [26] and Baccelli and Zuyev [5], they derive the optimal parameters to achieve minimum energy consumption. Experimental results for up to 10,000 nodes have been reported.

Power-aware virtual base stations (PA-VBS) [72, 74] is a first attempt to use residual power capacity to select the cluster heads in mobile ad hoc networks. It is attractive to WSNs because of its characteristics of load balancing and scalability to the growth of network size. A load-balanced clustering approach for heterogeneous sensor networks is introduced in Gupta and Younis [28]. The gateway nodes (cluster heads) with high energy manage the cluster member nodes and forward the data collected from the cluster member to the base station, which may be far from member nodes.

Table 5.3 provides a comparison of several hierarchical protocols used for E²IC in WSNs.

5.5.2.2   Energy-Efficient Information Dissemination (E²ID) Protocols

Information dissemination plays a critical role in WSNs. This is particularly the case in reactive and hybrid application scenarios, in which time-sensitive information should reach other nodes as soon as serious phenomenon, e.g., early warning of a fire, is detected by some nodes, or when a query with certain attribute values should spread in the system in a timely manner. In general, information dissemination is conducted similarly to flooding, but conventional flooding schemes will cause problems of redundancy and overlap that lead to significant energy waste. In order to prolong the system lifetime, E²ID protocols are in great demand. Most E²ID apply the data aggregation function (discussed later) to eliminate redundant information.

A family of E²ID protocols named sensor protocols for information via negotiation (SPIN) have been proposed in Heinzelman et al. [31] and Julik et al. [48]. They employ a new type of control message — metadata to allow negotiation between neighboring nodes — so that a node only forwards a packet to a neighbor that wants to receive the data. In such a way, the energy waste caused by classical flooding schemes can be reduced. However, overhead of control messages is created for negotiation, which will lead to long latency. Moreover, each individual node must constantly maintain a neighbor list and update it periodically. This not only requires memory space, but will also cost extra energy. Therefore, generating and controlling metadata are critical to the success of the SPIN protocols family.

Directed diffusion [42, 44] incorporates in-network data aggregation, data caching, and data-centric dissemination while enforcing adaptation to the empirically best path. It aims to establish efficient *n*-way communication from single or multiple sources to sinks. Heidemann and colleagues [35] present a physical implementation of directed diffusion with a wireless sensor test bed and show that the traffic can be reduced by up to 42% when deploying a duplicate suppression data aggregation method (see Section 5.5.3.1).

Another E^2ID scheme is proposed in Ye et al. [99]. Instead of generating a control message to implement data aggregation as other E^2ID proposals do, it attempts to relieve redundant information by allowing immediate nodes to conduct a random back-off procedure to delay the packet delivery. During this deferral period, if an intermediate node receives new data from other source nodes, it will combine them with the previous ones then transmit the processed data. Therefore, it trades latency for energy saving.

5.5.3   Signal and Data Processing

As has been mentioned in previous subsections, most E^2 communication protocols incorporate localized signal and data processing to reduce the amount of traffic in the network. Two main categories of signal and data processing applied in WSNs are (1) data aggregation and (2) collaborative signal and information processing (CSIP).

5.5.3.1   Data Aggregation

The principle of data aggregation or data fusion is to minimize traffic load (in terms of number and/or length of packets) by eliminating redundancy. It applies a novel data-centric approach to replace the traditional address-centric approach in data forwarding [47]. Specifically, when an intermediate node receives data from multiple source nodes, instead of forwarding all of them directly, it checks the contents of incoming data and then combines them by eliminating redundant information under the constraints of acceptable accuracy.

Several data aggregation algorithms have been reported in the literature. The most straightforward is duplicate suppression, i.e., if multiple sources send the same data, the intermediate node will only forward one of them. Using a maximum or minimum function is also possible. Heinzelman and colleagues [31] and Julik and colleagues [48] proposed SPIN (see Section 5.5.2.2) to realize traffic reduction for information dissemination using metadata negotiations between sensors to avoid redundant and/or unnecessary data propagation through the network. The greedy aggregation approach [43] can improve path sharing and attain significant energy savings when the network has higher node densities compared with the opportunistic approach.

Krishnamachari and colleagues [47] described the impact of source–destination placement on the energy costs and delay associated with data aggregation. They also investigated the complexity of optimal data aggregation. In Reference 111, a polynomial-time algorithm for near-optimal maximum lifetime data aggregation (MLDA) is described for data collection in WSNs. The scheme is superior to others in terms of systems lifetime, but has a high computational expense for large sensor networks. A simple and efficient clustering-based heuristic for maximum lifetime data aggregation (CMLDA) has been proposed by Dasgupta and coworkers [17] for small- and large-scale sensor networks.

5.5.3.2   Collaborative Signal and Information Processing (CSIP)

CSIP schemes are also powerful in reducing the amount of traffic transmitted and thus result in energy efficiency in WSNs. With the combination of interdisciplinary techniques, such as low-power communication and computation; space–time signal processing; distributed and fault-tolerant algorithms; adaptive systems; and sensor fusion and decision theory, CSIP is expected to provide solutions to many challenges. These include dense spatial sampling of interested events; distributed asynchronous processing; progressive accuracy; optimized processing and communication; data fusion; querying; and routing tasks [49].

CSIP can be implemented through coherent signal processing on a small number of nodes in a cluster or through noncoherent processing across a larger number of nodes when synchronization is not a strict requirement [23]. CSIP algorithms can be classified [67] as information-driven schemes [101, 102];

mobile agent-based schemes [66], which attempt to reduce the system traffic by employing an agent, thus transmitting the integration process (code) to the data sites instead of moving original data directly; and relation-based schemes [108], which use a top-down approach to select the sensor nodes to sense and communicate based on a high-level description of the task.

5.6   Concluding Remarks

Wireless sensor networks, which incorporate the functions of sensing; date collection and storage; computation and processing; communication through wireless medium; and/or actuating, have been envisioned for a wide spectrum of applications in various military and civil domains. Because of their great potential for providing safer and healthier environments for human beings through ubiquitous monitoring, objective localization, and target tracking, they have attracted extensive interest from industry and academia.

However, due to the tiny size of individual sensor nodes and human-unattended operation manners, WSNs often encounter severe resource constraints — especially, limited power supply. Therefore, from a networking perspective, energy efficiency is a critical objective in the design of communication protocols. Several proposals have been made in this direction, including physical layer approaches; data link and MAC layer protocols; network layer protocols; and transport layer and application layer strategies, as well as energy-efficient software development. This chapter has provided a classification and comparative study of such proposals.

Extending the lifetime of the network is the major concern of all energy-efficiency protocols. In order to gain a constant power supply, various proposals have been put forward to substitute conventional batteries in WSNs. With advances in power techniques, more energy-efficient, environmentally friendly alternative energy sources are raising research interest. The most familiar is called green power — solar technologies. During full and bright sunlight hours, solar cells can provide up to 45 mW/in.^{2} [112]. However, because of the variable nature of solar energy and the lack of cost-effective electricity storage techniques, solar power has been unable to become the sole power supply and is unlikely to grab a large share of the energy generation market.

Fuel cells offer a promising solution to WSNs' energy challenges. It is estimated that fuel cell use will grow by a factor of 250 over the next decade [113]. These "cells operate much like batteries, converting chemical energy into electrical power supply; however, unlike batteries, they never run down or require recharging" [114]. For example, the recent invention of regenerative fuel cells and zinc air fuel cells can be operated as a "closed loop" system in which no additional external fuel is necessary. Although they are currently just in the research and development stage, fuel cells are a potential low-cost, high-power replacement to the existing battery in WSNs. Another interesting scheme is to make sensor nodes self-contained and self-powered — able to harvest energy supply from the environment such as vibrations caused by motor vehicles driving down nearby streets or people walking on raised floors. Power output between tens and hundreds of microwatts per cubic centimeter is possible from vibrations in a normal office building with current MEMS technology [11, 81, 115].

Certainly, future WSNs can be equipped with more than one power supply. For example, a solar cell and battery can work together. The battery is used to initiate the solar cell's activity, or act as a standby power resource when there is not enough sunlight to make the solar cell create the required power. Alternatively, two power supplies can be assigned dynamically to drive distinct components of the sensor node respectively according to their residual energy.

In summary, new energy sources can have a significant impact on the development of WSNs, including [113]:

- Improving the efficiency of energy systems
- Ensuring longevity against energy disruption
- Expanding future energy choices
- Promoting energy production and use in ways that respect health and environmental values

References

1. J. Agre and L. Clare, An integrated architecture for cooperative sensing networks, *Computer Mag.*, 33(5), 106–108, 2000.
2. I.F. Akyildiz, W. Su, Y. Sankarasubramaniam, and E. Cayirci, Wireless sensor network: a survey, *Computer Networks*, 38(4), 393–422, March 2002.
3. G. Anastasi, L. Lenzini, and E. Mingozzi, Stability and performance analysis of HIPERLAN, *IEEE INFOCOM 1998*, 1, 134–141, San Francisco, 1998.
4. G. Asada et al., Wireless integrated network sensors: low power systems on a chip, *24th IEEE Eur. Solid-State Circuits Conf.*, 9–12, The Hague, the Netherlands, 1998.
5. F. Baccelli and S. Zuyev, Poisson Voronoi spanning trees with applications to the optimization of communication networks, *Operations Res.*, 47(4), 619–631, 1999.
6. D.J. Baker, and A. Ephremides, The architectural organization of a mobile radio network via a distributed algorithm, *IEEE Trans. Commun.*, 29(11), 1694–1701, November 1981.
7. S. Bandyopadhyay and E.J. Coyle, An energy efficient hierarchical clustering algorithm for wireless sensor networks, *IEEE INFOCOM 2003*, 3, 1713–1723, San Francisco, 2003.
8. L. Benini and G.D. Micheli, *Dynamic Power Management: Design Techniques and CAD Tools*, Kluwer Academic Pub., Norwell, MA, 1998.
9. D.W. Carman, P.S. Kruus, and B.J. Matt, Constraints and approaches for distributed sensor network security, NAI labs technical report 00-010, September 2000.
10. A. Cerpa, J. Elson, D. Estrin, L. Girod, M. Hamilton, and J. Zhao, Habitat monitoring: application driver for wireless communications technology, *1st ACM SIGCOMM Workshop Data Commun. Latin Am. Caribbean*, 31(2), 20–41, San Jose, Costa Rica, April 2001.
11. L. Chandrakasan et al. Design considerations for distributed microsensor systems, *IEEE 1999 Custom Integrated Circuits Conf.*, 279–286, San Diego, CA, May 1999.
12. K. Chakrabarty, S.S. Iyengar, H. Qi, and E. Cho, Coding theory framework for target location in distributed sensor networks, *Int. Symp. Inf. Technol.: Coding Computing*, 130–134, Las Vegas, 2001.
13. K. Chakrabarty, S.S. Iyengar, H. Qi, and E. Cho, Grid coverage for sureillance and target location in distributed sensor networks, *IEEE Trans. Computers*, 51, 1448–1453, 2002.
14. B. Chen et al., SPAN: An energy-efficient coordination algorithm for topology maintenance in ad hoc wireless networks, *ACM/IEEE MOBICOM 2001*, 85–96, Rome, 2001.
15. T. Clouqueur, V. Phipatanasuphorn, P. Ramanathan, and K.K. Saluja, Sensor deployment strategy for target detection, *ACM WSNA*, 42–48, Atlanta, 2002.
16. Collaborative Sensor Networks, Internet article, http://wwwhome.cs.utwente.nl/~havinga/sensor.html, 2003.
17. K. Dasgupta, K. Kalpakis, and P. Namjoshi, An efficient clustering-based heuristic for data gathering and aggregation in sensor networks, *IEEE WCNC'03*, 3, 1948–1953, New Orleans, 2003.
18. J. Deng, and Z.J. Hass, Dual busy tone multiple access (DBTMA): a new medium access control for packet radio networks, *IEEE ICUPC'98*, 2, 973–977, Florence, 1998.
19. S.S. Dhillon, K. Chakrabarty, and S.S. Iyengar, Sensor placement for grid coverage under imprecise detections, *Int. Conf. Inf. Fusion (FUSION) 2002*, 2, 1581–1587, Annapolis, 2002.
20. S. Dulman et al., Collaborative communication protocols for wireless sensor networks, *Eur. Res. Middleware Architectures for Complex Embedded Syst. Workshop*, Pisa, 2003.
21. M. Easton, Using space technology to fight malaria, *Queen's Gazette*, April 7, 2003, 13.
22. D. Estrin, R. Govindan, J. Heidemann, and S. Kumar, Next century challenges: scalable coordination in sensor networks, *IEEE MOBICOM 1999*, 263–270, August 1999.
23. D. Estrin, L. Girod, G. Pottie, and M. Srivastava, Instrumenting the world with wireless sensor networks, *IEEE ICASSP 2001*, 4, 2033–2036, 2001.
24. ETSI, HIPERLAN Functional Specification, ETSI draft standard, July 1995, http://www.etsi.org.
25. J. Feng, F. Koushanfar, and M. Potkonjak, System architectures for sensor networks issues, alternatives, and directions, *IEEE ICCD'02: VLSI Computers Processors*, 226–231, Freiburg, Germany, 2002.

26. S.G. Foss and S.A. Zuyev, On a Voronoi aggregative process related to a bivariate poisson process, *Adv. Appl. Probability*, 28(4), 965–981, 1981.

27. A.J. Goldsmith and S.B. Wicker, Design challenges for energy-constrained ad hoc wireless networks, *IEEE Wireless Commun.*, 8–27, August 2002.

28. G. Gupta and M. Younis, Load-balanced clustering of wireless sensor networks, *IEEE ICC 2003*, 3, 1848–1852, 2003.

29. V. Gutnik and A.P. Chandrakasan, An embedded power supply for low-power DSP, *IEEE Trans. VLSI Syst.*, 5(4), 425–435, December 1997.

30. P.J.M. Havinga and G.J.M. Smit, Energy-efficient wireless networking for multimedia applications, *Wireless Communications and Mobile Computing*, Wiley, 165–184, 2001.

31. W. Heinzelman, J. Kulik, and H. Balakrishnan, Adaptive protocols for information dissemination in wireless sensor networks, *ACM MOBICOM 1999*, 174–185, Seattle, 1999.

32. W. Heinzelman, A. Chandrakasan, and H. Balakrishnan, Energy-efficient communication protocol for wireless microsensor networks, *HICSS 2000*, Maui, 8020–8029, January 2000.

33. W. Heinzelman, Application-specific protocol architecture for wireless networks, Ph.D. dissertation, Massachusetts Institute of Technology, June 2000.

34. W.B. Heinzelman, A.P. Chandrakasan, and H. Balakrishnan, An application-specific protocol architecture for wireless microsensor networks, *IEEE Trans. Wireless Commun.*, 1(4), October 2002, 660–670.

35. J. Heidemann et al., Building efficient wireless sensor networks with low-level naming, *18th ACM Symp. Operating Syst. Principles*, 146–159, Banff, October 2001.

36. J. Hill et al., System architecture directions for networked sensor networks, *9th Int. Conf. Architectural Support Programming Languages Operating Syst.*, 28(5), 93–104, Cambridge, MA, November 2000.

37. J. Hill and D. Culler, A wireless embedded sensor architecture for system level optimization, University of California Berkeley Technical Report, 2002.

38. J. Hill et al., TinyOS: operating system for sensor networks, hppt://tinyos.millennium.berkeley.edu, 2003.

39. X. Hong, M. Gerla, H. Wang, and L. Clare, Load balanced, energy-aware communications for Mars sensor networks, *IEEE Aerospace Conf.*, 3, 1109–1115, 2002.

40. A. Howard, M.J. Mataric, and G.S. Sukhatme, Mobile sensor network deployment using potential fields: a distributed, scalable, solution to the area coverage problem, *6th Int. Symp. Distributed Autonomous Robotics Syst.* (DAR02) Fukuoka, Japan, 299–308, June 2002.

41. IEEE 802.11 WG, Information technology — telecommunications and information exchange between systems. Local and metropolitan area networks — specific requirements. Part 11: wireless LAN medium access control (MAC) and physical layer (PHY) specifications, 1997.

42. C. Intanagonwiwat, R. Govindan, and D. Estrin, Directed diffusion: a scalable and robust communication paradigm for sensor networks, *ACM/IEEE MOBICOM 2000*, 56–67, Boston, August 2000.

43. C. Intanagonwiwat, D. Estrin, and R. Govindan, Impact of network density on data aggregation in wireless sensor networks, technical report 01-750, University of Southern California, November 2001.

44. C. Intanagonwiwat et al. Directed diffusion for wireless sensor networking, IEEE/ACM Trans. Networking, 11(1), 2–16, February 2003.

45. J.M. Kahn, R.H. Katz, and K.S.J. Pister, Next century challenges: mobile networking for Smart Dust, *ACM MOBICOM*, 271–278, Seattle, 1999.

46. F. Koushanfar, M. Potkonjak, and A. Sangiovanni–Vincentelli, Fault-tolerance techniques for sensor networks, *IEEE Sensors*, 2, 1491–1496, 2002.

47. B. Krishnamachari, D. Estrin, and S. Wicker, Impact of data aggregation in wireless sensor networks, *Int. Workshop Data Aggregation Wireless Sensor Networks*, 575–578, Vienna, Austria, July 2002.

48. J. Julik, W. Heinzelman, and H. Balakrishnan, Negotiation-based protocols for disseminating information in wireless sensor networks, *Wireless Networks*, 8, 169–185, 2002.

49. S. Kumar, F. Zhao, and D. Shepherd, Collaborative signal and information processing in microsensor networks, *IEEE Signal Processing Mag.*, 13–14, March 2002.

50. J. F. Kurose, and K. W. Ross, *Computer Networking, a Top-Down Approach Featuring the Internet*, 1st ed., Addison–Wesley, Longman, MA, 2000.

51. S. Lindsey and C.S. Raghavendra, PEGASIS: power-efficient gathering in sensor information systems, *IEEE Aerospace Conf. 2002*, 3, 1125–1130, March 2002.

52. S. Lindsey, C. Raghavendra, and K.M. Sivalingam, Data gathering algorithms in sensor networks using energy metrics, *IEEE Trans. Parallel Distributed Syst.*, 13(9), 924–935, September 2002.

53. A. Mainwaring, J. Polastre, R. Szewczyk, D. Culler, and J. Anderson, Wireless sensor networks for habitat monitoring, *ACM WSNA'02*, 88–97, Atlanta, September, 2002.

54. A. Manjeshwar, and D. P. Agrawal, TEEN: a routing protocol for enhanced efficiency in wireless sensor networks, *IEEE IPDPS 2002 Workshop*, 2009–2015, April 2002.

55. A. Manjeshwar and D.P. Agrawal, APTEEN: a hybrid protocol for efficient routing and comprehensive information retrieval in wireless sensor networks, *IEEE IPDPS 2002 Workshop*, 195–202, April 2002.

56. A. Manjeshwar, Q. Zeng, and D.P. Agrawal, An analytical model for information retrieval in wireless sensor networks using enhanced APTEEN protocol, *IEEE Trans. Parallel Distributed Syst.*, 13(12), 1290–1302, December 2002.

57. S. Meguerdichian, F. Koushanfar, M. Potkonjak, and M.B. Srivastava, Coverage problems in wireless ad-hoc sensor networks, *IEEE INFOCOM*, 3, 1380–1387, Anchorage, 2001.

58. R. Min et al., Low-power wireless sensor networks, *IEEE VLSID 2001*, 205–210, India, 2001.

59. R. Min et al., Energy-centric enabling technologies for wireless sensor networks, *IEEE Wireless Commun.*, 28–39, August 2002.

60. J. Mirkovic, G.P. Venkataramani, S. Lu, and L. Zhang, A self-organizing approach to data forwarding in large-scale sensor networks, *IEEE ICC*, 5, 1357–1361, St. Petersburg, Russia, 2001.

61. S. Musman, P.E. Lehner, and C. Elsaesser, Sensor planning for elusive targets, *J. Computer Math. Modeling*, 25(3), 103–115, 1997.

62. A.K. Parekh, Selecting routers in ad-hoc wireless networks, *Proc. ITS*, 1994.

63. T.A. Pering, T.D. Burd, and R.W. Brodersen, The simulation and evaluation of dynamic voltage scaling algorithms, *Int. Symp. Low Power Electron. Design* (ISLPED), 1998.

64. G.J. Pottie and L.P. Clare, Wireless integrated network sensors: towards low cost and robust self-organizing security networks, *Proc. of SPIE 1998*, 3577, 86–95, 1999.

65. G.J. Pottie and W.J. Kaiser, Wireless integrated network sensors, *Commun. ACM*, 43(5), 51–58, 2000.

66. H. Qi, S.S. Iyengar, and K. Chakrabarty, Multi-resolution data integration using mobile agents in distributed sensor networks, *IEEE Trans. Syst., Man Cybernetics (part C)*, 31, 383–391, August 2001.

67. H. Qi, P.T. Kuruganti, and Y. Xu, The development of localized algorithm in wireless sensor networks, *Sensors Mag.*, 2, 286–293, 2002.

68. J.M. Rabaey et al., PicoRadio supports ad hoc ultra-low power wireless networking, *IEEE Computer Mag.*, 33(7), 42–48, 2000.

69. V. Raghunathan, C. Schurgers, S. Park, and M.B. Srivastava, Energy-aware wireless microsensor networks, *IEEE Signal Process. Mag.*, 40–50, 2000.

70. T.S. Rappoport, *Wireless Communications, Principles and Practice*, Prentice Hall, Upper Saddle River, NJ, 1996.

71. S. Ray et al., Robust location detection in emergency sensor networks, *IEEE INFOCOM 2003*, 2, 1044–1053, March 2003.

72. A. Safwat, H. Hassanein, and H.T. Mouftah, Power-aware fair infrastructure formation for wireless mobile ad hoc communications, *IEEE GLOBECOM 2001*, 5, 2822–2836, November 2001.

73. A. Safwat, H. Hassanein, and H. Mouftah, Optimal cross-layer designs for energy-efficient wireless ad hoc and sensor networks, *22nd IEEE Int. Performance, Computing, Commun. Conf.*, (IPCCC 2003), 123–128, April 2003.

74. A. Safwat, H. Hassanein, and H.T. Mouftah, Power-aware virtual base stations (PW-VBS) for wireless mobile ad hoc communications, *J. Computer Networks*, 41(3), 331–346, 2003.

75. A. Salhieh et al., Power efficient topologies for wireless sensor network, *Int. Conf. Parallel Process.*, 156–163, Spain, 2001.

76. L. Schwiebert, S.K.S. Gupta, and J. Weinamann, Research challenges in wireless networks of bio-medical sensors, *ACM MOBICOM 2001*, 151–165, Rome, July 2001.

77. C. Schurgers, V. Tsiatsis, and M. Srivastava, STEM: topology management for energy efficient sensor networks, *2002 IEEE Aerospace Conf.*, 3, 1099–1108, March 2002.

78. C. Schurgers, V. Tsiatsis, S. Ganeriwal, and M. Srivastava, Optimizing sensor networks in the energy–latency–density design space, *IEEE Trans. Mobile Computing*, 1(1), 70–80, January–March 2002.

79. C-C. Shen, C. Srisathapornphat, and C. Jaikaeo, Sensor information networking architecture and applications, *IEEE Personal Commun.*, 52–59, August 2001.

80. E. Shih et al., Physical layer driven protocol and algorithm design for energy-efficient wireless sensor networks, *ACM/IEEE MOBICOM'01*, 272–287, July 2001.

81. J.L. da Silva Jr. et al. Design methodology for PicoRadio networks, *Design, Automation Test Eur.*, DATE'2001, March 2001, Germany.

82. A. Sinha and A. Chandrakasan, Energy aware software, *VLSID'00*, January 2000.

83. A. Sinha and A. Chandrakasan, Dymamic power management in wireless sensor networks, *IEEE Design Test Computers*, 18(2), 62–74, 2001.

84. J.L. Sobrinho and A.S. Krishnakumar, Real-time traffic over the IEEE802.11 medium access control layer, *Bell Labs Tech. J.*, 172–187, Autumn 1996.

85. K. Sohrabi, J. Gao, V. Ailawadhi, and G.J. Pottie, Protocols for self-organization of a wireless sensor network, *IEEE Personal Commun.*, 16–27, October 2000.

86. M. Srivastava, R. Muntz, and M. Potkonjak, Smart kindergarten: sensor-based wireless networks for smart developmental problem-solving environments, *ACM MOBICOM 2001*, 132–138, Italy, 2001.

87. J.A. Stankovic et al., Real-time communication and coordination in embedded sensor networks, *Proc. IEEE*, 91(7), 1002–1022, 2003.

88. L. Subramanian and R.H. Katz, An architecture for building self-confirgurable systems, *MOBIHOC 2000*, 63–67, Boston, 2000.

89. A.S. Tanenbaum, *Computer Networks*, Prentice Hall, Upper Saddle River, NJ, 1996.

90. B. Tavil, and W.B. Heinzelman, TRACE: time reservation using adaptive control for energy efficiency, *IEEE JSAC*, 21(10), 1506–1515, 2003.

91. B. Tavil, and W.B. Heinzelman, MH-TRACE: multi-hop time reservation using adaptive control for energy efficiency, *MILCOM 2003*, Boston, 2003.

92. S. Tilak, N.B. Abu–Ghazaleh, and W. Heinzelman, Infrastructure trade-offs for sensor networks, *ACM WSNA'02*, 49–58, September 2002.

93. A. Wang and A. Chandrakasan, Energy efficient system partitioning for distributed wireless sensor networks, 2001 *IEEE Int. Conf. Acoustics, Speech, Signal Processing*, 2, 905–908, Salt Lake City, 2001.

94. A. Willig, R. Shah, J. Rabaey, and A. Wolisz, Altruists in the PicoRadio sensor network, *4th IEEE Int. Workshop Factory Commun. Syst.*, 175–184, Sweden, August 2002.

95. A.D. Wood and J.A. Stankovic, Denial of service in sensor networks, *IEEE Computer*, 35(10), 48–56, October 2002.

96. Y. Xu, J. Heidemann, and D. Estrin, Geography-informed energy conservation for ad hoc routing, *ACM MOBICOM 2001*, 70–84, Italy, 2001.

97. K. Xu, Performance analysis of differentiated QoS MAC in wireless area networks (WLANs), master degree thesis, Queen's University, 2003.
98. M.D. Yarvis et al., Real-world experiences with an interactive ad hoc sensor network, *Int. Conf. Parallel Processing Workshops (ICPPW'02)*, 2002.
99. F. Ye et al., A scalable solution to minimum cost forwarding in large sensor networks, *IEEE ICCCN*, 304–309, 2001.
100. W. Ye, J. Heidemann, and D. Estrin, An energy-efficient MAC protocol for wireless sensor networks, *IEEE INFOCOM*, 3, 1567–1576, New York, 2002.
101. F. Zhao, J. Shin, and J. Reich, Information-driven dynamic sensor collaboration for tracking applications, *IEEE Signal Process. Mag.*, 19(2), 61–72, March 2002.
102. F. Zhao et al., Collaborative signal and information processing: an information directed approach, *Proc. IEEE*, 91(8), 1199–1209, 2003.
103. Y. Zou and K. Chakrabarty, Sensor deployment and target localization based on virtual forces, *IEEE INFOCOM 2003*, 2, 1293–1303, March 2003.
104. Z.J. Hass, and J. Deng, Dual busy tone multiple access (DBTMA) — performance evaluation, *IEEE Vehicular Technol. Conf. (VTC) 1999*, May 1999.
105. H. Qi, S.S. Iyengar, and K. Chakrabarty, Distributed sensor networks: a review of recent research, *J. Franklin Inst.*, 338, 655–668, 2001.
106. K.S.J. Pister, SMART DUST: autonomous sensing and communication in a cubic millimeter, internet article, http://www-bsac.eecs.berkeley.edu/~pister/Smart/Dust/.
107. H.S. Carvalho, et al., A general data fusion architecture, *6th Intl. Conf. on Information Fusion (Fusion 2003)*, July 2003, Australia, 1465–1472.
108. L.J. Guibas, Sensing, tracking, and reasoning with relations, *IEEE Signal Processing Magazine*, 19(2), 73–85, March 2002.
109. J.-F. Frigon, V.C.M. Leung, and H.C.B. Chan, Dynamic reservation TDMA protocol for wireless ATM networks, *IEEE JSAC*, 19(2), 370–383, February 2001.
110. J.L. Sobrinho and A.S. Krishnakumar, Quality-of-service in ad hoc carrier sense multiple access networks, *IEEE JSAC*, 17(8), 1353–1968, August 1999.
111. K. Kalpakis, K. Dasgupta, and P. Namjoshi, Efficient algorithms for maximum lifetime data gathering and aggregation in wireless sensor networks, Technical Report UMBC-TR-02-13, Computer Science and Electrical Engineering Department, University of Maryland, Baltimore County, 2002.
112. How many solar cells would I need in order to provide all of the electricity that my house needs? internet article, http://science.howstuffworks.com/question418.htm.
113. Allied business intelligence, 2001, internet article, http://www.eyeforfuelcells.com.
114. Advancing fuel cells for clean and efficient power, internet article, Pacific Northwest National Laboratory, PNNL-SA-35945, February 2002, http://www.pnl. gov.
115. K.S.J. Pister, My view of sensor networks in 2010, internet article, http://robotics.eecs.berkeley.edu/~pister/SmartDust/in2010.

6

Fundamental Protocols to Gather Information in Wireless Sensor Networks

Jacir L. Bordim
Adaptive Communications Research Laboratories

Koji Nakano
Hiroshima University

6.1 Introduction

Advances in microelectronic mechanical systems (MEMS) and wireless communication technologies, and the availability of sophisticated sensor-signal processing algorithms have enabled production of multifunctional and small-sized sensor nodes [25, 29]. Due to their compactness and low cost, wireless sensors can be embedded and distributed at a fraction of the cost of conventional wired sensors. A wide range of tasks can be performed by these tiny devices, such as remote object monitoring and tracking; detection of the presence or absence of certain elements; and condition-based maintenance, among other special applications.

The physical world generates a vast amount of information that can be sensed/controlled. However, bandwidth and radio frequencies are finite resources. In addition, the energy cost for communications is generally much larger than the computational cost, which exacerbates the need to process raw data at the source and carefully control the access of the wireless medium. Although sensing devices are already used in a variety of applications, interconnecting them to perform a larger task has yet to become commonplace. Designing systems involving hundreds, or even thousands, of sensors will be a challenging task.

In general, sensors are battery-driven devices that operate on a limited energy budget. Furthermore, they must have a reasonable lifespan to be worth deploying. Clearly, in a network with thousands of sensors, battery replacement is not an option, thus requiring efficient management of the energy resources. Despite these limitations, there is a promising scope for wireless sensor networks (WSNs) in the near future [1, 15]. Indeed, sensor networks have been receiving increasing attention, not only from academia, but also from government and industry [27].

This chapter focuses on fundamental protocols for a large number of sensors in the context of WSNs [4, 9]. A WSN is a distributed system consisting of a base station and a number of wireless sensor nodes equipped with wireless radio communication capabilities. Energy consumption is a major factor in determining the lifespan of a sensor node. It is well known that a sensor utilizes a significant amount of energy while sending or receiving packets. Power dissipation is also expressive even when a sensor receives a packet not intended for it. State-of-the-art systems support a number of power states of the sensor nodes (see Reference 28). To simplify matters, here the assumption is that each sensor node in a WSN has two power states: *awake* and *asleep*. Although energy dissipation is negligible in the asleep mode, a significant amount of energy is utilized in the awake state. In MICA2's processor/radio transceiver, for instance, a sensor consumes up to 27 mA to send and receive packets, and less than 1 μA while asleep [28].

In this chapter, the efficiency of a protocol will be assessed by two metrics: (1) the overall amount of time required by the protocol to terminate; and (2) for each individual sensor node, the total amount of time it must be awake to transmit/receive packets. The goals of optimizing these parameters are, of course, conflicting. Sometimes, one can easily minimize the overall completion time at the expense of energy consumption and vice versa. The challenge is to strike a sensible balance between the two by designing protocols that are time and energy efficient.

This chapter first presents efficient collaborative computations and communications strategies to solve a number of fundamental problems in the context of WSNs. More specifically, it shows a number of efficient protocols to aggregate and process information among the sensor nodes and to make such information available at the base station. The fundamental problems considered in this chapter are *information gathering* and *faulty node location*.

In many applications, the sensor nodes must aggregate the sensed/monitored data. Because the sensor nodes are empowered with the ability to share their observations and coordinate among themselves to gather and process information, meaningful information can be transferred to the base station. Such information can then be retrieved and used to control the environment from remote locations.

Presented first are energy-efficient protocols that compute the sum of n numbers over any commutative and associative binary operator stored in n wireless sensor nodes arranged in a two-dimensional grid of size $\sqrt{n} \times \sqrt{n}$. This begins with a protocol that computes the sum in $O\left(r^2 + \left(\dfrac{n}{r^2}\right)^{\frac{1}{3}}\right)$ time slots with no sensor node awake for more than $O(1)$ time slots, where r is the transmission range of the sensor nodes. Then a fault-tolerant protocol that computes the sum in the same number of time slots with no sensor node awake for more than $O(\log r)$ time slots is presented. Finally, it is shown that, in a WSN where the sensor nodes are empowered with the ability to adjust their transmission range r dynamically during the execution of the protocol, the sum can be computed in $O(\log n)$ time slots and no sensor node needs to wake for more than $O(\log n)$ time slots.

A sensor node may cease its sensing task due to power dissipation or when affected by external events. As the number of faulty sensor nodes in the WSN increases, the accuracy of the sensed/monitored data is likely to deteriorate. If the state of the sensors in the network is known, new sensors can be added to affected areas in order to regain the desired degree of accuracy.

The second major topic in this chapter is to design efficient protocols to identify and locate the state of the sensor nodes in the WSN. As before, consider a WSN populated by n wireless sensor nodes arranged in a two-dimensional grid of size $\sqrt{n} \times \sqrt{n}$. Let q and k denote the number of fault-free and faulty nodes in the WSN, respectively. It will be shown that the task of identifying the faulty nodes and reporting their location to the base station can be completed, with high probability, in $O(\alpha + r^2)$ time slots and none of the sensors needs to wake for more than $O(\log \log \alpha)$ time slots, where $\alpha = \min(q, k)$ and r is the transmission range of the sensor nodes.

Section 6.2 in this chapter gives a formal description of the model. The main results are presented in Section 6.3 and Section 6.4. Section 6.3 begins by presenting some preliminary results that are later used in developing energy-efficient protocols to gather information on a WSN. Section 6.4 presents an energy- and time-efficient protocol to solve the detection of faulty nodes in a WSN.

6.2   Model Definition

The sensor nodes in a WSN are tiny devices operating on batteries and employing low-power radio transceivers to enable communication. The base station is equipped with a powerful antenna that enables it to monitor all the sensor nodes under consideration. It is assumed that the amount of power necessary for a sensor node to communicate with the base station does not exceed the amount of power necessary to communicate with neighboring sensor nodes. This is a reasonable assumption because one can increase the base station's receiver front-end sensitivity [24].

Another possibility is to couple the sensor nodes with *corner cube reflectors* technology to enable passive communications [14, 26, 29]. It is assumed that the base station and all the sensor nodes have a local clock that keeps synchronous time, perhaps by interfacing with the base station or through a GPS. All sensor nodes run the same protocol and can perform computations on the data being sensed. As is customary, time is assumed to be slotted and all transmissions take place at slotted boundaries [3, 6]. It is assumed that at any time slot, a sensor node can communicate with the base station and vice versa. The size of a data packet is such that its transmission can be completed within one time slot.

In a *single-hop* WSN, a sensor node can directly communicate with any other sensor node. In a *multihop* WSN, however, the communication between two sensor nodes may involve a sequence of hops through a chain of pairwise adjacent sensor nodes. There is a single-hop communication between the base station and the sensor nodes, while the communication among the sensor nodes can be single or multihop. There are several possible models for WSNs; this chapter considers WSNs in which all the sensor nodes in the network are fixed, homogeneous, and energy constrained.

A commonly accepted assumption is employed: when two or more sensor nodes are in transmission range of each other and transmitting in the same time slot, the corresponding packets *collide* and are garbled beyond recognition. Similarly, when two or more sensor nodes are broadcasting a packet in the same time slot, the base station cannot receive these packets. The computation among the sensor nodes is performed in coordination with the base station. A sensor node in a single-hop WSN can tune to a channel to send/receive a packet. At the end of a time slot, the status of the channel can be:

- NULL: no packet has been driven into the channel in the current time slot
- SINGLE: exactly one packet has been driven into the channel in the current time slot
- COLLISION: two or more packets have been driven into the channel in the current time slot

Suppose that a sensor node is positioned in a two-dimensional plane. When a sensor node transmits a packet with power r, the signal will be strong enough for other sensor nodes to hear it within the Euclidean distance r from the sensor node that originates the packet. In other words, to cover a range of r, the sensor node that originates the signal must transmit with enough power to cover that range. Every sensor node in the *intensity zone*, that is, the region within the distance r from a sensor node that originates the packet, is guaranteed to receive it. It is well known that signals are subject to fluctuations and start fading after traveling some distance [20]. Thus, sensors outside the transmission range r of a source node, e.g., $r + \delta$ for some $\delta > 0$, may or may not receive the packet.

This situation is formalized as follows: the *fading zone* of a sensor is defined as the region outside the *intensity zone* and inside the circle with radius $f(r)$, where f is an increasing function. Those sensor nodes in the *fading zone* may or may not receive the packet. The status of the channel is always SINGLE in the *intensity zone*, whereas in the *fading zone*, it is SINGLE or NULL. The sensors in the *silent zone*, that is, beyond the Euclidean distance $f(r)$ from the sensor that originated the broadcast, are guaranteed not to receive the packet, and the status of the channel is always NULL. Figure 6.1 depicts the transmission zones of a sensor node as described here.

For simplicity, assume that $f(r) = 2r$ and design the protocols under this assumption. These protocols work for any general function f as long as $f(r) = c \cdot r + o(r)$, for any fixed $c \geq 1$, by adjusting some

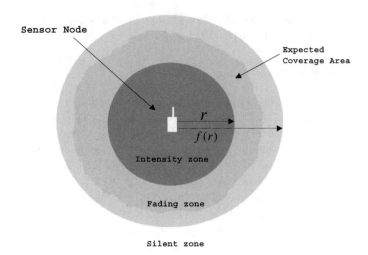

FIGURE 6.1 Transmission zones of a sensor node.

parameters used in the protocols. Although the signal may attenuate in the presence of objects, all sensor nodes in the transmission range of r are ensured to hear from the sensor that originates the signal.

Observe the channel status of a sensor node. For this purpose, let D be a sensor node in a WSN and let S be the unique sensor node broadcasting in a given time slot. The channel status of D is NULL only if D lies in the *silent zone* of S. Otherwise, the channel status is SINGLE if D lies in the *intensity zone* of S and either SINGLE or NULL if D is in the *fading zone*. Now, consider the case in which two or more sensor nodes are broadcasting at the same time. Clearly, if their transmissions do not interfere (i.e., do not overlap), the channel status of D is as discussed here.

In case of overlapping transmissions, the channel status is as follows. When D lies in the *fading zones* of two or more sensor nodes, the channel status can be NULL, SINGLE, or COLLISION. The channel status is SINGLE or COLLISION when D lies in the *intensity zone* of one sensor and in the *fading zone(s)* of other sensor(s). The channel status is COLLISION when D lies in the *intensity zones* of two or more sensors. Therefore, a sensor node is ensured to receive a packet, only if it lies in the *intensity zone* of the source node and no interference occurs from other broadcasts. Figure 6.2 illustrates the channel status in which the transmissions of two sensor nodes overlap.

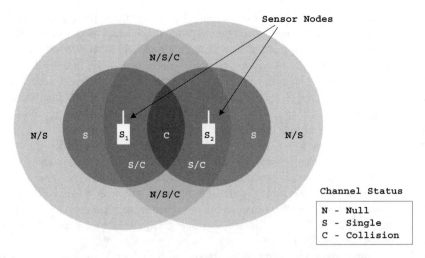

FIGURE 6.2 Status of the channel when the transmissions of two sensor nodes diverge.

In this chapter, it is assumed that the sensor nodes in the WSN are organized as a two-dimensional square plane of size $\sqrt{n} \times \sqrt{n}$ with coordinates (x,y), $(1 \leq x, y \leq \sqrt{n})$. The plane can be viewed as n cells of unit size 1×1. Let $C(x,y)$, $(1 \leq x, y \leq \sqrt{n})$, denote a cell consisting of all points (x',y'), $(x \leq x' < x + 1; y \leq y' < y + 1)$. Suppose that each cell $C(x,y)$ has a sensor denoted as $S_{x,y}$. Throughout this chapter it is assumed that each sensor node $S_{x,y}$, $(1 \leq x, y \leq \sqrt{n})$ knows its cell's location, that is, integers x and y.

If the sensors on a WSN of size $\sqrt{n} \times \sqrt{n}$ can broadcast with sufficient power to cover an area of $\sqrt{2}n$, then any pair of sensors can directly communicate. In such cases, the WSN essentially allows a single-hop communication. In other words, if the sensors with transmission range r are allocated on a WSN of size $\dfrac{\sqrt{2}r}{2} \times \dfrac{\sqrt{2}r}{2}$ ($\approx 0.71r \times 0.71r$), then a single-hop communication is ensured.

The next section discusses a fundamental protocol to collect information in WSNs and, following that, an energy-efficient protocol to identify faulty nodes in a WSN.

6.3   Gathering Information in Wireless Sensor Networks

One of the salient features of WSNs is information gathering, the ultimate goal of which is to group and collect the information sensed by the sensor nodes. This section deals with a protocol to assemble and retrieve such information. More specifically, time- and energy-efficient protocols are proposed to compute the sum over any commutative and associative binary operator for the values stored in the sensor nodes. The binary operators can be addition; multiplication; logical AND/OR; finding the maximum/minimum; etc.

Because the sensor node may fail due to power depletion or intentional or unintentional damages, a protocol has been devised that works even in the presence of faulty sensor nodes. In developing fault-tolerant protocols, it is assumed that at least one sensor remains fault free for a given group during the course of the protocol.

In the following, a formal problem definition is given. For this purpose, consider a WSN in which each sensor, $S_{i,j}$, $(1 \leq i, j \leq \sqrt{n})$, has a value $x_{i,j}$. Such value could represent temperature, humidity, gravity, seismic information, etc. Let $\otimes$ be any commutative and associative binary operator, such as addition, multiplication, or finding the maximum. The *sum* problem is to perform the $\otimes$ operation over all $x_{i,j}$, that is,

$$X = x_{1,1} \otimes x_{1,2} \otimes \cdots \otimes x_{\sqrt{n},\sqrt{n}}$$

The next subsection presents some preliminary results to solve the *sum* problem on single-hop WSNs. These results will be used later in developing protocols for multihop WSNs.

6.3.1   Preliminaries

Consider a single-hop WSN comprising m sensor nodes, in which each sensor has a unique ID in $[1,m]$. Let S_i denote a sensor node with ID i, $(1 \leq i \leq m)$, which has a value x_i stored in it. In this scenario, the sum problem can be solved in $m - 1$ time slots as follows: for each time slot i, $(1 \leq i \leq m - 1)$, the sensor node S_i broadcasts x_i on the channel and sensor node S_{i+1} monitors the channel to receive x_i. Then, S_{i+1} computes $x_{i+1} = x_i \otimes x_{i+1}$. After $m - 1$ iterations, node S_m holds the overall sum. Note that this protocol is energy efficient because each sensor node is awake for, at most, two time slots. For later reference, consider the following simple result:

Lemma 6.1. The sum problem can be solved in a single-hop WSN comprising m sensor nodes in m − 1 *time slots with each sensor node awake for, at most, two time slots.*

In case of node failure, this protocol is unable to yield the correct result. However, a way to overcome this situation is presented next.

In developing a fault-tolerant protocol, assume that faults do not occur during the execution of the protocol and at least one sensor node remains active, i.e., fault free, in the WSN. For each time slot $i \left(1 \leq i \leq m\right)$, sensor node S_i broadcasts x_i on the channel, and every sensor node $S_k, (k > i)$ monitors the channel to receive the value. After receiving x_i, each sensor node S_k computes $x_k = x_i \otimes x_k$. When a sensor node $S_j (j < i)$ hears the broadcast of S_i, it knows that S_i exists and leaves the protocol. Consequently, this protocol terminates in m time slots. The worst case occurs when a single sensor node remains active in the WSN. In such a case, this sensor node will be awake for exactly m time slots. Thus:

Lemma 6.2. Even in the presence of faulty sensor nodes, the sum can be computed on a WSN with m *sensor nodes in* m *time slots and no sensor node is awake for more than* m *time slots.*

Although this protocol is fault tolerant, it is not energy efficient. Now a fault-tolerant and energy-efficient protocol is introduced to compute the sum in a single-hop WSN. When the protocol terminates, the following two conditions are satisfied:

  A. The last awake sensor node, denoted as S_k, such that no sensor node S_i with $i > k$ exists, is identified and holds the final result.
  B. The protocol takes $2m - 2$ time slots and no sensor node is awake for more than $2\log m$ time slots.

If $m = 1$, then S_1 knows x_1 and conditions A and B are verified. Now, assume that $m \geq 2$. The m sensor nodes are partitioned into two groups $S_1 = \left\{ S_i \middle| 1 \leq i \leq \dfrac{m}{2} \right\}$ and $S_2 = \left\{ S_i \middle| \dfrac{m}{2} + 1 \leq i \leq m \right\}$. Recursively compute the sum in P_1 and P_2. By the induction hypothesis, the A and B conditions are satisfied and therefore each of the two subproblems can be solved in $m - 2$ time slots, with no sensor node awake for more than $2\log m - 2$ time slots.

Let S_j and S_k be the last active sensor nodes in groups P_1 and P_2, respectively. In the next time slot, sensor node S_j transmits the sum $\Sigma\{x_i \,|\, 1 \leq i \leq j \text{ and } S_i \text{ exists}\}$ on the channel. The last active sensor node, S_k, in P_2 monitors the channel and updates the result. In one additional time slot, the sensor node S_k reveals its identity. The reader can easily confirm that the protocol satisfies the aforementioned conditions A and B. The following lemma summarizes the preceding discussion:

Lemma 6.3. Even in the presence of faulty sensor nodes, the sum can be computed on a WSN with m *sensor nodes in* 2m − 2 *time slots with no sensor node awake for more than* 2log m *time slots.*

6.3.2   Protocols to Solve the Sum Problem in Multihop WSNs

Because each sensor node $S_{i,j}, \left(1 \leq i, j \leq \sqrt{n}\right)$ knows its cell's location within the WSN, a naive protocol can compute the sum in $O(n)$ time slots as follows: each sensor node $S_{i,j}$ broadcasts, one at a time, its value $x_{i,j}$ on the channel. The base station monitors the channel and computes the final result. Clearly, this approach is energy efficient because each sensor node is awake for only one time slot. However, it is not time efficient. The goal of this section is to present protocols that minimize the overall completion time while allowing the sensors to power-off their transceivers for the largest possible extent so as to save energy. First, an energy-efficient protocol is presented that solves the sum problem in $O\left(r^2 + \left(\dfrac{n}{r^2} \right)^{\frac{1}{3}} \right)$ time slots and the sensor nodes need only wake for a constant number of time slots.

6.3.2.1   Energy-Efficient Summing Protocol

The protocol begins by partitioning the n cells (nodes) into groups, blocks, and sub-blocks. Next, the sum is computed in a bottom-up fashion, starting with sub-blocks, then blocks, and finally groups. The sum of each group is later transmitted to the base station, which computes the overall sum. The parameter k used in the protocol determines the size of the groups. Later, in the description of the protocol, it will be shown how to set this parameter properly. The details of the protocol are given next.

PROTOCOL WSN_SUM

Step 1. The n cells are divided into n/k_2 groups of size $k \times k$. Each group is then partitioned into $k^2 / \left(\dfrac{9r}{4}\right)^2$ blocks of size $\dfrac{9r}{4} \times \dfrac{9r}{4}$. Each block is further partitioned into 81 sub-blocks of size $\dfrac{r}{4} \times \dfrac{r}{4}$.

Step 2. Compute the sum on each block.

Step 3. Compute the sum on each group by combining the partial results of Step 2.

Step 4. For each group, the sensor node that holds the sum of its group broadcasts it to the base station, one at a time. The base station monitors the channel and computes the overall sum.

The partitioning scheme in Step 1 allows for single hop within a sub-block as well as between any two sensor nodes lying on adjacent sub-blocks. The partitioning scheme is illustrated in Figure 6.3. Because each sensor node $S_{i,j}, \left(1 \le i, j \le \dfrac{9r}{4}\right)$ knows its location, this information can be easily converted into an ID number. Because each node has a unique ID number, the sum can be computed in top-down fashion, from left to right in odd rows and from right to left in even rows, using the protocol of Lemma 6.1. In other words, the sum on each block is computed in a snake-like fashion. Because the sensor nodes in adjacent blocks lie completely outside the transmission range of each other, Step 2 can be performed in parallel on all blocks. Thus, Step 2 can be computed in $O(r^2)$ time slots and no sensor node needs to wake for more than two time slots. Also, the sensor node that holds the sum of its block must wake for only one time slot.

Let $S'_{i,j}, \left(1 \le i, j \le \dfrac{r}{4}\right)$ denote a sensor node within a sub-block and $S''_{i,j}, \left(1 \le i, j \le \dfrac{k}{\frac{9r}{4}}\right)$ be the sensor node on the bottom right of each block that holds the sum of its corresponding block. Note that there are eight sub-blocks of size $\dfrac{r}{4} \times \dfrac{r}{4}$ between $S''_{i,j}$ and $S''_{i,j+1}$. Clearly, these two sensor nodes cannot directly communicate because they are outside the transmission range of each other. Therefore, sensors in the sub-blocks that lie between $S''_{i,j}$ and $S''_{i,j+1}$ will be used to forward packets from $S''_{i,j}$ to $S''_{i,j+1}$. In Step 3, when a sensor node, e.g., $S''_{i,j}$, broadcasts its value, the sensor node $S'_{\frac{r}{4},\frac{r}{4}}$ located in the right neighboring sub-block receives the packet and forwards it. This process is repeated until the value reaches the sensor node $S''_{i,j+1}$, which updates its value and broadcasts it in the next time slot, as illustrated in Figure 6.4.

When this process finishes, the sensor node in the rightmost column will hold the sum of its row. Next, the sum of the rightmost column is computed so that the sensor node located on the bottom-right corner of each block will hold the sum of its block. Computing the sum on the rows of all groups takes $9 \times \left(\dfrac{k}{\frac{9r}{4}} - 1\right)$ time slots; to compute the sum on the rightmost column requires additional $9 \times \left(\dfrac{k}{\frac{9r}{4}} - 1\right)$ time slots. Thus, Step 3 can be computed in $O\left(\dfrac{k}{r}\right)$ time slots; no sensor node needs to wake for more than three time slots and the sensor that holds the sum wakes for only two time slots.

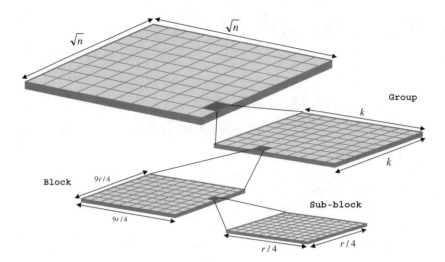

FIGURE 6.3 Grid partitioning scheme.

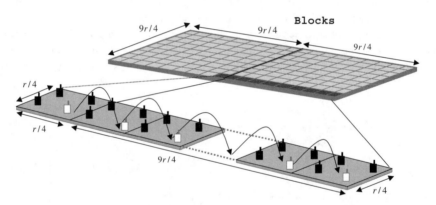

FIGURE 6.4 The computation in Step 3 is performed in parallel for every row on each group.

In Step 4, the sensor node located at the bottom-right corner of each group broadcasts the sum of its group to the base station. The base station monitors the channel and computes the overall sum. Because there are $\frac{n}{k^2}$ groups, this step takes $O\left(\frac{n}{k^2}\right)$ time slots to compute the sum and the sensor nodes must wake for exactly one time slot. Therefore, this protocol takes $O\left(r^2 + \frac{k}{r} + \frac{n}{k^2}\right)$ time slots to compute the sum of n numbers on a WSN with no sensor node awake for more than five time slots. The time complexity can be minimized by properly selecting the parameter k. With $k = (nr)^{\frac{1}{3}}$, the time complexity becomes $O\left(r^2 + \left(\frac{n}{r^2}\right)^{\frac{1}{3}}\right)$. Thus

Theorem 6.1. On a WSN in which the sensor nodes are arranged in cells of a grid of size $\sqrt{n} \times \sqrt{n}$, one sensor node per cell, the sum of n *numbers can be computed by an energy-efficient protocol in* $O\left(r^2 + \left(\dfrac{n}{r^2}\right)^{\frac{1}{3}}\right)$ *time slots when* r >> *1, and no sensor needs to wake for more than O(1) time slots.*

Furthermore, if the sensor nodes were able to adjust their transmission range before the execution of the protocol, r should be selected so that $r^2 = \left(\dfrac{n}{r^2}\right)^{\frac{1}{3}}$ in order to minimize the running time slots. If this is the case, $r = n^{\frac{1}{8}}$, and the time complexity of the preceding protocol becomes $O\left(n^{\frac{1}{4}}\right)$. Because this protocol relies on the non-fault-tolerant approach of Lemma 6.1, it may not yield the correct result in the presence of faulty nodes. The next sub-section presents a fault-tolerant protocol for multihop WSNs.

6.3.2.2 Fault-Tolerant Energy-Efficient Summing Protocol

In developing a fault-tolerant and energy-efficient summing protocol, it is assumed that at least one sensor node remains fault free in each sub-block during the course of the protocol. Here, Step 2 of the previous protocol is modified as follows:

PROTOCOL FAULT-TOLERANT WSN_SUM

Step 2.1. Compute the sum on each sub-block.
Step 2.2. Combine the partial sums of Step 2.1 to obtain the sum on each block.

After partitioning the grid, Step 2.1 computes the sum on each sub-block using the energy-efficient and fault-tolerant protocol of Lemma 6.3. Because Step 2.1 can be computed in parallel for neighboring blocks, it takes $O(r^2)$ time slots to compute the sum on each sub-block and no sensor needs to wake for more than $O(\log r)$ time slots. Let $S'_{i,j}$ $(1 \le i, j \le 9)$, be the sensor node that holds the sum of sub-block i,j at the end of Step 2.1. The sum of each block in Step 2.2 is computed in a snake-like fashion by combining the partial results of Step 2.1. Because there are 81 sub-blocks, the sum on blocks can be computed in $O(1)$ time slots and no sensor node needs to wake for more than two time slots. Step 4 and Step 5 are performed as in the previous protocol.

Overall, the fault-tolerant protocol takes $O\left(r^2 + \dfrac{k}{r} + \dfrac{n}{k^2}\right)$ time slots to compute the sum of n numbers and no sensor node needs to wake for more than $O(\log r)$ time slots. By selecting $k = (nr)^{\frac{1}{3}}$, the time complexity becomes $O\left(r^2 + \left(\dfrac{n}{r^2}\right)^{\frac{1}{3}}\right)$. To summarize:

Theorem 6.2. On a WSN in which the sensor nodes are arranged in cells of grid size $\sqrt{n} \times \sqrt{n}$, one sensor node per cell, the sum of n *numbers can be computed by a fault-tolerant and energy-efficient protocol in time slots* $O\left(r^2 + \left(\frac{n}{r^2}\right)^{\frac{1}{3}}\right)$ *when* r >> 1, *and no sensor needs to wake for more than* $O(\log r)$ *time slots.*

When the sensor nodes are empowered with the ability to change their transmission range, then with $r = n^{\frac{1}{8}}$ and the time complexity of the preceding protocol becomes $O\left(n^{\frac{1}{4}}\right)$.

6.3.3   WSNs with Dynamic Transmission Range

This section presents a protocol to solve the sum problem in WSNs in which the sensors have the ability to adjust their transmission range during the course of the protocol. Because the size of the blocks and sub-blocks dynamically changes during the execution, transmission range must change accordingly to ensure connectivity with neighboring sensors in the same sub-block. The details of the protocol are spelled out as follows:

PROTOCOL DYNAMIC WSN_SUM;

---

for $i = 1$ to $\lceil \log \sqrt{n} \rceil$ **odo**

  $k \leftarrow 4(2^i)$;

  every sensor sets its transmission range r to $2^i \sqrt{2}$;

  divide the $\sqrt{n} \times \sqrt{n}$ grid into $\dfrac{n}{k^2}$ blocks of size $k \times k$;

  divide each $k \times k$ block into $\dfrac{k^2}{\left(2^i\right)^2}$ sub-blocks of size $2^i \times 2^i$;

    for $l = 1$ to $\dfrac{n}{k^2}$ **do in parallel** for each block $k \times k$

    for $j = 1$ to 16 do
        compute the *sum* on each sub-block j;
      endfor
    endfor
  endfor

In each iteration of the outermost **for**-loop, the transmission range is properly chosen so that the communication among the sensors within a sub-block is ensured. The grid partitioning used here is similar to that of the previous protocols. In the first iteration, the size of each block is 8×8 and the size of each sub-block is 2×2. Note that each sub-block contains four cells of sensors and each block has 16 sub-blocks. When the size of a sub-block is $2^i \times 2^i$ the transmission range is set to $2^i \sqrt{2}$. This ensures communication among sensors within the same sub-block. Because the distance between any two sub-blocks, located in different blocks, is at least $3.5(2^i)$, the sum on neighboring blocks can be performed in parallel.

The communication among the sensors within the same sub-block is guaranteed, so the sum can be computed sequentially on each sub-block j, $(1 \leq j \leq 16)$ using Lemma 6.1 or Lemma 6.3. Using Lemma 6.1, the sum on each sub-block takes exactly four time slots. Thus, computing the sum on all sub-blocks takes 64 time slots. Note that only sensors that hold the sum of its sub-block in iteration $i - 1$ will participate in iteration i. Each sub-block on iteration i is formed by four sub-blocks of the previous iteration. Then, by adjusting the transmission range on iteration i, the sum on sub-blocks can be computed as discussed in the first iteration.

The outermost **for**-loop repeats for $\lceil \log \sqrt{n} \rceil$ times, and the innermost **for**-loop can be computed in constant time. Hence, our algorithm runs in $64 \cdot \frac{1}{2} \lceil \log n \rceil$ time slots using Lemma 6.1. Note that the sensor holding the sum, at the end of the protocol, is located at the bottom-right corner of the grid, and it has been awake for the largest amount of time. Specifically, the sensor that holds the sum at the end of the protocol has been awake for $2 \cdot \frac{1}{2} \lceil \log n \rceil$ time slots, according to Lemma 6.1. With Lemma 6.3, the sum on each sub-block takes six time slots; in this case, the algorithm runs in $96 \cdot \frac{1}{2} \lceil \log n \rceil$ time slots and no sensor must be awake for more than $4 \cdot \frac{1}{2} \lceil \log n \rceil$ time slots. Partitioning the grid and recursively combining the partial results results in the protocol yielding the overall sum in $O(\log n)$ time slots using Lemma 6.1 or Lemma 6.3. The following theorem summarizes these findings:

Theorem 6.3. On a WSN grid of sensor nodes, where the sensors are endowed with the ability to select the desired transmission range, the sum problem over n *numbers can be computed in* O(log n) *time slots with no sensor node awake for more than* O(log n) *time slots.*

6.4 Identifying Faulty Nodes in Wireless Sensor Networks

A sensor node in a WSN may cease its sensing task due to power depletion or because it was affected or destroyed by external events. In favorable circumstances, a neighboring sensor may be able to cover, even partially, the sensing task of its neighboring faulty nodes. However, the accuracy of the sensed data tends to decrease as the number of faulty sensor nodes increases — to a point at which the sensed data may not correctly reflect the physical events. If the state of the sensors in the network is known, then faulty sensors could be repaired or new sensors added to those affected areas. This task can be carried on until the desired degree of coverage is obtained.

The task of identifying faulty nodes is fundamental in network design and in multiprocessor systems and has been extensively studied in the past [2, 5, 23]. However, these studies focus on wired networks, where energy consumption and limited channel capacity is not an issue. Recently, Chessa and Santi [12] proposed a faulty identification protocol for wireless sensor networks. They focused on "soft" faults, in which a faulty node continues to operate with an alternated behavior. Later, they proposed a protocol that constructs a tree; the information obtained by the leaf nodes is routed to a sink node [13].

Here, the focus is on the task of identifying faulty nodes in a WSN. Consider a WSN with n sensor nodes, where a number of these sensors have been affected by some external event that prevents them from continuing their sensing tasks. Let k denote the number of faulty nodes, and q $(= n - k)$ denote the number of fault-free nodes in the WSN. The *fault-location* problem is to identify the location of the k-faulty sensors in the WSN. In identifying faulty nodes, it is assumed that faults do not occur during the course of the protocol and that faults are permanent (i.e., a faulty node remains in that state until it is repaired or replaced). It is also assumed that the base station stores a $\sqrt{n} \times \sqrt{n}$ Boolean matrix, $B = (b_{ij})$, where each entry b_{ij}, $(1 \leq i, j \leq \sqrt{n})$, is associated with the state of the sensor node $S_{i,j}$.

In this scenario, one can easily solve the fault-location problem as follows: each sensor node S_i, $(1 \leq i \leq n)$, broadcasts one at a time and the base station monitors the channel to check whether the sensor is faulty or fault-free. After n time slots, the base station knows the position within the grid of each fault-free node and, consequently, it can determine the position of the faulty nodes. This intuitive approach is energy efficient because each sensor node must wake for only one time slot. However, it is not time efficient.

Next, a time- and energy-efficient protocol that identifies and reports the location of faulty nodes to the base station is presented. It will be shown that the fault-location task can be performed in $O(\alpha + r^2)$ time slots and none of the sensors needs to wake for more than $O(\log \log \alpha)$ time slots, where $\alpha = \min(q,k)$ and r is the transmission range of the sensor nodes.

6.4.1 Preliminaries

This section presents some preliminary results that will be used in the subsequent section. To begin, suppose that a WSN is populated by n identical sensors that cannot be distinguished by any serial or manufacturing number. The *initialization* task is to assign a unique ID number in the range $[1,n]$ to each sensor so that no two sensors are assigned the same ID. To solve the initialization task, some protocols assume that the number n of sensors is known prior to the beginning of the protocol [22].

Of particular interest are the results obtained by Bordim and colleagues [10], in which the initialization task is carried out even if the number n of stations is unknown beforehand. These authors showed that a single-channel, single-hop ad hoc radio network (ARN) populated by n stations can be initialized with high probability in $O(n)$ time slots, with each station awake for $O(\log n)$ time slots, at most. More recently, they have shown that initialization can be done in the same number of time slots with each station awake for, at most, $O(\log \log n)$ time slots [11].

An ARN is a distributed system consisting of a number of wireless mobile stations that achieve communication without the aid of any network infrastructure or centralized administration. The main differences between WSNs and ARNs are the presence of a base station in the former, and the fact that the stations are mobile in the ARN. For later reference, the following important result is reproduced from Bordim et al. [11]:

Theorem 6.4. Even if the number n *of stations is not known beforehand, with probability exceeding 1 –* $O(n^{-1.5})$, *an* n-*station, single-channel, single-hop ARN can be initialized in* $O(n)$ *time slots, with no station awake for more than* $O(\log \log$ n$)$ *time slots.*

A single-hop ARN can be simulated in two time slots by a multihop WSN by relaying packets to the base station. To see this, suppose that sensor S wants to send a packet to sensor D, where S and D are outside radio reach of each other. In the first time slot, sensor S broadcasts its packet to the base station. In an additional time slot, the base station broadcasts the packet and sensor D wakes to receive it. Because a single-hop communication is between the base station and the sensor nodes in the WSN, a packet can be routed from a source sensor to any destination sensor in the WSN in two time slots — that is, one time slot to route to the base station and another for the base station to send the packet to its final destination. For the latter's reference, consider the following corollary:

Corollary 6.1. Let n *denote the number of sensors in a multihop WSN. Even if the number* n *of sensors is not known beforehand, with probability exceeding* $1 - O(n^{-1.5})$, *the* n *sensors, can be initialized in* $O(n)$ *time slots, with no sensor awake for more than* $O(\log \log$ n$)$ *time slots.*

6.4.2 Locating Faulty Sensors in Multihop WSNs

The main contribution of this section is to present an energy- and time-efficient protocol to solve the faulty location problem in multihop WSNs. As a preliminary step, the multihop WSN is partitioned into groups and the position of the faulty nodes within each group is obtained. Next, each sensor node with faulty nodes in its vicinity sends a packet containing their location, along with its own location, to the base station. Upon learning the location of the faulty nodes, the base station can identify the position of each faulty node as well as the position of each fault-free node within the WSN. The details of this fault-location protocol for multihop WSNs follow:

PROTOCOL FAULT LOCATION

Step 1. The $\sqrt{n} \times \sqrt{n}$ cells are divided into $\dfrac{n}{9r^2}$ groups, with each group containing $3r \times 3r$ cells.

Step 2. The sensor nodes in each group learn the state, faulty or fault free, of its immediate neighbors.

Step 3. Each sensor node with faulty sensors in its vicinity informs the base station of its location and the location of its neighboring faulty nodes,

---

Obviously, the partitioning scheme of Step 1 requires no broadcast. After partitioning, the fault-location task is carried out in each group $G_l, \left(1 \le l \le \dfrac{n}{9r^2}\right)$. Let $S^l_{i,j}, (1 \le i, j \le 3r)$denote a sensor node within group G_l. Initially, each sensor $S^l_{i,j}$ learns the state, faulty or fault free, of its immediate neighboring sensors located at its north, south, east and west positions. This is achieved by having each sensor node $S^l_{i,j}$ transmitting on the channel and the sensor nodes $S^l_{i-1,j}, S^l_{i+1,j}, S^l_{i,j-1}$, and $S^l_{i,j+1,j}$ monitoring the channel. Because the coordinates can be easily converted into a unique ID number, each sensor knows the exact time slot in which it must broadcast. Similarly, each sensor knows when it must be awake to monitor the channel. By checking the channel status at the appropriate time, the immediate neighbors of $S^l_{i,j}$ know that $S^l_{i,j}$ is fault free, if the channel status is SINGLE, and faulty if the channel status is NULL.

It is not difficult to see that a sensor node located at the boundary of a group can learn the status of its adjacent neighboring sensors located in a neighboring group by listening to the channel at the appropriate time slot. Clearly, each sensor node must wake for, at most, five time slots (one time slot to broadcast on the channel) so that its neighbors learn its state. At most, it must wake for four time slots to record the state of its immediate neighbors.

To avoid collision within each group G_l, only one sensor is allowed to broadcast at a time. Therefore, $9r^2$ time slots are necessary for each sensor node to learn the state of its four immediate neighbors. Note that the sensor nodes in the corresponding locations of any two adjacent groups lie completely outside the transmission range of each other because the minimum distance between them is greater than $2r$. Thus, one can reuse the channel and compute Step 2 in parallel for neighboring groups without incurring collisions. Therefore, Step 2 can be performed in $O(r2)$ time slots.

Let P denote the set of sensors that have identified faulty nodes in their vicinity at the end of Step 2. The next task is to have each sensor node in P communicating the location of its faulty neighbors to the base station. To avoid collision there, only one sensor is allowed to route its items to the base station at a time. Thus, each sensor node in P must learn the exact time slot in which it can wake and route its packet to the base station so that no other sensor is transmitting at the same time. The idea is to assign unique IDs to each sensor in P. Remember that these IDs are temporary and should not be confused with the sensors' permanent IDs. Once these IDs have been assigned, each sensor can route its packet to the base station in $|P|$ time slots without collision at the base station.

Let p denote the number of sensors in P, that is, $p = |P|$. As the exact number of sensors in P is unknown, we cannot rely on initialization protocols that require such information. Corollary 6.1 states that even if the number p of sensors is unknown beforehand, the p sensors can still be initialized with high probability. More precisely, according to this corollary, the task of assigning a unique ID number in the range $[1,p]$ to each sensor in P so that no two sensors have been assigned the same ID can be performed in $O(p)$ time slots with no sensor node in P awake for more than $O(\log \log p)$ time slots with high probability.

Let $S_m, (1 \le m \le p)$ so that $S_m \in P$. After completing the initialization task, each sensor S_m wakes at time slot m and routes its packet to the base station. Because each sensor has, at most, four faulty neighbors, this information can be sent in a single packet containing their respective locations. The base station monitors the channel and collects the packets that are routed for it. Clearly, this can be done in p time slots. Thus, Step 3 can be computed in $O(p)$ time slots. The task of identifying the faulty nodes and reporting their locations to the base station can be completed in $O(p + r2)$ time slots and none of the sensors need wake for more than $O(\log \log p)$ time slots with high probability.

Now consider some special cases:

- All the sensors have crashed (i.e., $k = n$).
- A single fault-free sensor is left in the field (i.e., $k = n - 1$).
- All the sensors are fault free (i.e., $k = 0$).

Verifying whether $k = n$ takes only one time slot and can be achieved as follows. All the sensors broadcast their locations and the base station monitors the channel. If the channel status is NULL, the base station learns that all the sensors have crashed and the protocol terminates. Similarly, if the channel status is SINGLE, the base station knows the location of the unique fault-free sensor in the WSN and the protocol finishes. In case of COLLISION, at least two fault-free nodes must be left in the field. However, the base station cannot check whether $k = 0$ or not. For this purpose, after Step 2, all sensors that have faulty nodes in their vicinity broadcast and the base station monitors the channel. If the channel status is NULL, then all the sensors are fault free and the protocol terminates. The following lemma summarizes the results obtained so far.

Lemma 6.4. Let p *denote the number of sensors in a multihop WSN that have identified faulty neighbors adjacent to it. The task of identifying the faulty nodes and reporting them to the base station can be completed, with high probability, in* $O(p + r2)$ *time slots and none of the sensors need wake for more than* $O(\log \log p)$ *time slots.*

As shown in Figure 6.5(a), the maximum number of sensors reporting to the base station occurs when $k = n/2$. Clearly, if the number of faulty sensors increases beyond $n/2$, the number of sensors reporting to the base stations decreases. On the other hand, when the number k of faulty sensors decreases, so does the number of sensors reporting to the base station because only sensors with faulty neighbors in their vicinity report to the base station (see Figure 6.5b). Thus, when the number of faulty sensors is small, at most, $4k$ sensors report to the base station. In other words, for large k, $p \le n - k$ and, for small k, $p \le 4k$.

Let q denote the number of fault-free nodes in the WSN. Clearly, the number of sensor nodes reporting to the base station cannot surpass q. As discussed earlier, for a small number of faulty nodes, $q > 4k$. Thus, in the presence of k faulty nodes, at most, $\min(q, 4k)$ sensors report to the base station. The following theorem summarizes this discussion.

Theorem 6.5. Let q *and* k *denote the number of fault-free and faulty nodes in the WSN, respectively. The task of identifying the faulty nodes and reporting their location to the base station can be completed, with high probability, in* $O(\alpha + r^2)$ *time slots and none of the sensors need wake for more than* $O(\log \log \alpha)$ *time slots, where* $\alpha = \min(q, k)$, *and* r *is the transmission range of the sensor nodes.*

Once the location of the fault sensors is obtained, the base station can identify the areas populated by faulty nodes and the areas populated by fault-free nodes. This is a trivial task when the location of all the faulty nodes is reported to the base station as in Figure 6.5(a). Clearly, the base station can identify "boundary" locations with the information received from the reporting nodes. As shown in Figure 6.5(b) and (c), such boundary locations consist of reporting nodes on one side and faulty nodes on the other. By checking such boundaries, the base station can determine the location of the faulty nodes; that is, the base station can identify which side is populated by faulty nodes and which is populated by fault-free nodes. It should be clear at this point that the base station can determine the status of each sensor node in the WSN from the information provided by the reporting nodes.

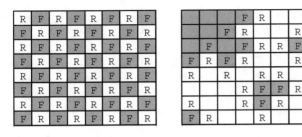

FIGURE 6.5 (a) Worst-case scenario in which half of the sensors report to the base station; (b) and (c) depict boundary formations and identify the reporting nodes and location of the faulty nodes (reported faulty nodes) as viewed by the base station.

6.5 Conclusions

WSNs can greatly augment the ability to control and supervise the environment from distant locations. For reliable monitoring, in many situations the sensor nodes must be in close proximity to the physical events. This may require deployment of large numbers of such devices, which, in turn, demands efficient distributed algorithms to enable the sensor nodes to operate in coordination with other sensors to collect and process data. By sharing observations, the sensor nodes can gather relevant data and transfer meaningful information to a sink node.

This chapter focused on the design of efficient collaborative computation and communication strategies to solve a number of fundamental problems in the context of WSNs. It began by discussing a number of energy-efficient protocols to compute the sum of n numbers over any commutative and associative binary operator stored in n wireless sensor nodes arranged in a two-dimensional grid of size $\sqrt{n} \times \sqrt{n}$. It was shown that the sum can be computed in $O\left(r^2 + \left(\dfrac{n}{r^2}\right)^{\frac{1}{3}}\right)$ time slots with no sensor node awake for more than $O(1)$ time slots, where r is the transmission range of the sensor nodes.

Next, discussion focused on a fault-tolerant protocol that computes the sum in the same number of time slots with no sensor node awake for more than $O(\log r)$ time slots. Then, it was demonstrated that in WSNs in which sensor nodes are empowered with the ability to adjust their transmission range, the sum can be computed in $O(\log n)$ time slots and no sensor node needs to wake for more than $O(\log n)$ time slots. Finally, a time- and energy-efficient protocol to identify the state of the sensor nodes in the WSN was presented. Here, the task of identifying faulty nodes and reporting their location to the base station can be completed, with high probability, in $O(\alpha + r^2)$ time slots and none of the sensors need wake for more than $O(\log \log \alpha)$ time slots, where q and k denote the number of fault-free and faulty nodes in the WSN, respectively; $\alpha = \min(q,k)$; and r is the transmission range of the sensor nodes.

References

1. I.F. Akyildiz, W. Su, Y. Sankarasubramaniam, and E. Cayirci, Wireless sensor networks: a survey, *Computer Networks*, 38 393–422, 2002.
2. A. Bagchi and S.L. Hakimi, An optimal algorithm for distributed system level diagnosis, *Proc. FTCS-21*, 214– 221, 1991.

3. R. Bar–Yehuda, O. Goldreich, and A. Itai, Efficient emulation of single-hop radio network with collision detection on multi-hop radio network with no collision detection, *Distributed Computing*, 5, 67–71, 1991.

4. R.S. Bhuvaneswaran, J.L. Bordim, J. Cui, and K. Nakano. Fundamental protocols for wireless sensor networks, *IEEE Trans. Fundamentals*, E-85A(11), 2479–2488, Nov. 2002.

5. D.M. Blough and H.W. Wang, The broadcast comparison model for on-line fault diagnosis in multicomputer systems: theory and implementation, *IEEE Trans. Computers*, 48(5), 470–493, May 1999.

6. D. Bertzekas and R. Gallager, *Data Networks*, 2nd ed., Prentice Hall, Upper Saddle River, NJ, 1992.

7. R. Binder, N. Abramson, F. Kuo, A. Okinaka, and D. Wax, ALOHA packet broadcasting — a retrospect, *AFIPS Conf. Proc.*, May 1975, 203–216.

8. U. Black, *Mobile and Wireless Networks*, Prentice Hall, Upper Saddle River, NJ, 1996.

9. J.L. Bordim, F. Hsu, and K. Nakano, Identifying faulty nodes in wireless sensor networks, *J. Interconnection Networks*, 3(3 & 4), 197–211, 2002.

10. J.L. Bordim, J. Cui, T. Hayashi, K. Nakano, and S. Olariu, Energy-efficient initialization protocols for ad-hoc radio networks, *IEEE*, E83-A(9), 1796–1803, Sep. 2000.

11. J.L. Bordim, J. Cui, N. Ishii, and K. Nakano, Doubly logarithmic energy-efficient initialization protocols for single-hop radio networks, *IEEE Trans. Fundamentals*, E83-A(9), 1796–1803, Sep. 2000.

12. S. Chessa and P. Santi, Comparison-based system level fault diagnosis in ad hoc networks, Proc. *IEEE 20th Symp. Reliable Distributed Syst.* (SRDS), New Orleans, 257–266, October 2001

13. S. Chessa and P. Santi, Crash faults identification in wireless sensor networks, *Computer Commun.*, 25(14), 1273–1282, Sept. 2002.

14. P.B. Chu, N.R. Lo, E. Berg, and K.S.J. Pister, Optical communication using micro corner cube reflectors, *10th IEEE Int. Micro Electro Mechanical Syst. Conf.* (MEMS 97), Nagoya, Japan, Jan. 26–30, 1997, 350–355.

15. D. Estrin, R. Govindan, J. Heidemann, and S. Kumar, Next century challenges: scalable coordination in sensor networks. In *Proc. 5th Annu. Int. Conf. Mobile Computing Networks* (MobiCOM'99), Seattle, WA, August 1999.

16. K. Feher, *Wireless Digital Communications*, Prentice Hall, Upper Saddle River, NJ, 1995.

17. W.C. Fifer and F.J. Bruno, Low-cost packet radio, *Proc. IEEE*, 75, 33–42, 1987.

18. V.K. Garg and J E. Wilkes, *Wireless and Personal Communication Systems*, Prentice Hall, Englewood Cliffs, NJ, 1996.

19. M. Gerla and T.-C. Tsai, Multicluster, mobile, multimedia radio network, *Wireless Networks*, 1, 255–265, 1995.

20. M.P.M. Hall and L.W. Barclay *Radio-Wave Propagation*, IEEE Electro Magnetic Wave Series 30, Peter Peregrinus Ltd., 1989.

21. E.P. Harris and K.W. Warren, Low-power technologies: a system perspective, *Proc. 3rd Int. Workshop Multimedia Commun.*, Princeton, 1996.

22. T. Hayashi, K. Nakano, and S. Olariu, Randomized initialization protocols for packet radio networks, *Proc. 13th Int. Parallel Process. Symp.*, (1999), 544–548, 1999.

23. S. Hosseini, J. Kuhl, and S. Reddy, *A Diagnosis Algorithm for Distributed Computing Systems with Dynamic Failure and Repair*, IEEE Trans. Computers, 33(3), 223–233, Mar. 1984.

24. J.C. Liberty and T. Rappaport, *Smart Antennas for Wireless Communications: IS-95 and Third Generation CDMA Applications*, Prentice Hall PTR, Upper Saddle River, NJ, 1999.

25. T.-H. Lin, H. Sanchez, H.O. Marcy, and W.J. Kaiser, Wireless integrated network sensors (wins) for tactical information systems, in *Proc. 1998 Government Microcircuit Applications Conf.*

26. V.S. Hsu, MEMS corner cube retro-reflectors for free-space optical communications: research project, University of California, Berkeley, 1999.

27. G.J. Pottie, Wireless integrated network sensors (WINS): the Web gets physical, *Bridge* 31(4), Winter 2001.
28. MICA2 Motes, http://www.xbow.com/.
29. B. Warneke, M. Last, B. Leibowitz, and K.S.J. Pister, Smart Dust: communicating with a cubic-millimeter computer, *IEEE Computer Mag.*, 44–51, 2001.

7

Design Challenges in Energy-Efficient Medium Access Control for Wireless Sensor Networks

Duminda Dewasurendra
Virginia Tech

Amitabh Mishra
Virginia Tech

7.1 Introduction

Wireless sensor networks (WSNs) are an emerging paradigm posing new challenges for researchers in wireless communications [1]. This new class of networks closely resembles the behavior of wireless ad hoc networks. Nevertheless, they have a few unique differences; the principal one is the small size of nodes constituting a WSN. Although smaller nodes make WSNs suitable for several existing and emerging applications related to information sensing, this also implies that the nodes have limited resources, i.e., CPU speed, memory, battery, and radio interface. Because the nodes are resource constrained, they require network designs that can be customized for different types of application environments, thus placing significant demands on algorithm design, protocol specification, and technologies.

This chapter focuses on medium access control (MAC) schemes for WSNs. Unique features of these networks will be briefly discussed in order to highlight the issues demanding special attention during the design of MAC schemes. Significant research efforts currently underway in this context will be studied along with MAC schemes for generic wireless ad hoc networks (WAHNs) and wireless local area networks (WLANs). Finally, the challenges and open issues related to MAC algorithm design for the effective deployment of future WSNs will be discussed.

7.2 Unique Characteristics of Wireless Sensor Networks

WSNs consist of large numbers of distributed nodes that organize themselves to form a multihop wireless network. Each node consists of one or more types of sensors, an embedded processor, small memory, and a low-power radio transceiver. Generally, these nodes are battery powered and coordinate among themselves to achieve a common task. Compared to nodes in a generic WAHN operating under IEEE 802.11 [2] or Bluetooth [3, 4] protocols, these nodes are extremely small in size and possess limited energy resources. The transmitting power and thus the communication range are much lower, which is largely compensated by a higher density of nodes in most cases. WSNs can have distributed, hierarchical, or clustered architectures, as illustrated in Figure 7.1.

The lack of centralized control is common to WSNs as well as to other WAHNs. Nevertheless, the behavior of a WSN is largely governed by the application for which it is used. Even considering a single application, the desired role of nodes would be different from time to time. For example, in a battlefield application, it may be employed to monitor the ambient data patterns silently and generate alarms if the specified deviations are observed. The same network may be used to track the movement of a detected vehicle at another time. Such dynamic changes of network objectives and the corresponding change in node behavior are uncommon in most of the other generic WAHNs.

Furthermore, nodes of a sensor network are mostly unattended after deployment, permitting neither upgrade of energy sources nor troubleshooting. The node hardware is designed so that the overall cost is extremely low and nodes can be abandoned once the power sources are exhausted. Voids of discarded nodes may be filled with redundant nodes due to high node density, and perhaps by the deployment of additional nodes if the need arises. It is necessary that the network should accommodate such losses and new additions with least effort. These are unique issues pertaining to WSNs compared to generic WAHNs in which the nodes are mostly attended; energy sources are high capacity and can be recharged or replaced; and nodes have direct and individual interaction with users. Comparison between a WSN and a generic ad hoc network is summarized in Table 7.1.

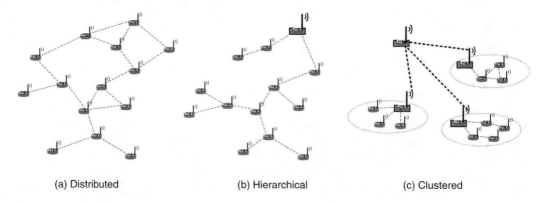

(a) Distributed   (b) Hierarchical   (c) Clustered

FIGURE 7.1 Sensor network architectures.

TABLE 7.1 Comparison of Features for WSNs and WAHNs

Wireless Sensor Network	Wireless ad hoc Network
Nodes involved in sensing the environment; events occurring in the environment can initiate certain communication in the network	No sensing behavior; network communication governed by user applications
Nodes are smaller in size	Larger nodes (e.g., PDAs, laptops)
Small and limited capacity power sources	High-capacity power sources
Inexpensive nodes	Relatively expensive nodes
Nodes unattended after deployment and designed for a prolonged lifetime with no maintenance or troubleshooting	Node troubleshooting and battery replacement possible
Node lifetime depends on the usage of attached power source	Node lifetime does not depend on energy resources because power sources are replaceable or rechargeable
Higher node density/highly redundant networks	Low node density/less redundant networks
Shorter transmission range (3 to 30 m)	Longer transmission range (10 to 500 m)
Limited processing and memory capacity	Higher processing power and memory
Nodes may stay in sleep mode for a significant amount of time	Nodes will be listening to the wireless medium most of the time
Data-centric communication; packet destination will depend on attributes of gathered data	Communication mostly occurs between specific nodes according to user requirements
Traffic profiles likely have statistically correlated properties comprising bursty traffic in case of event detection and low, continuous traffic during other times	Mostly continuous traffic, e.g., multimedia data streams
Low bandwidth (1–100 kb/s)	High bandwidth (e.g., 1 to 54 Mbps in IEEE 802.11-based WAHNs)
Network operation can be task oriented	Operation similar for all applications

7.2.1 Why Are MAC Layer Design Issues Important?

In all wireless networks, nodes must share a single medium for communication. Network performance largely depends upon how efficiently and fairly the nodes can share this common medium. Note that the packet transmission is directly handled by the MAC layer. Compared to a wired medium, a significant portion of the node's energy is spent on radio transmissions and on listening to the medium for anticipated packet reception. On the other hand, wireless networks always have restricted power sources; thus, careful design of the MAC scheme is necessary for the optimal performance and extended lifetime of the network.

In the context of WSNs, this requirement is extremely critical. According to the characteristics highlighted previously, nodes of a WSN carry extremely low energy resources and remain unattended after deployment; therefore, the node lifetime depends entirely on how energy is conserved during communication. Although some exhausted nodes could be compensated using redundant neighboring nodes, certain situations may arise rendering a part of the network completely inactive due to low connectivity and insufficient coverage, or making that part of the network inaccessible and isolated from the other parts. Such scenarios could be averted by avoiding unnecessary transmissions and longer listening periods — activities that consume the highest amount of power in nodes.

Another related issue is the high node density in WSNs. Although the transmission ranges are lower, a fairly high number of nodes can contend for the medium, at least in certain portions of the network. By the same token, transmissions from each node would increase the background noise for a large number of nodes, which may disrupt their own receptions. Thus, the MAC schemes for WSNs should be carefully designed to achieve the optimum performance toward the intended application. Previous surveys [1, 5] discuss some issues related to medium access in WSNs and WAHNs.

7.3   MAC Protocols for Wireless Ad Hoc Networks

The closest types of networks rendering a similar behavior to WSNs are WAHNs, although they have marked differences as highlighted in our discussion. Properly standardized MAC protocols designed to cater to the ad hoc and distributed nature of WAHNs have been developed and are in commercial use. Also, some of them focus on energy savings, mainly for mobile applications. These features are highly sought after in WSNs as well.

Currently available MAC protocols for wireless ad hoc networks are of two major types: *contention based* (CSMA) and *scheduling based* (TDMA, FDMA, or CDMA). In contention-based MAC schemes, the nodes compete among each other for channel access, whereas in scheduling-based methods, a specific schedule of channel access is used in time, frequency, or code domains. This section will briefly discuss several important medium access schemes belonging to both categories, including IEEE 802.11 [2]; Bluetooth [3, 4]; energy-conserving MAC (EC-MAC) [6]; and the power aware multiple access (PAMAS) [7]. Their merits, drawbacks, and suitability for WSNs will be highlighted.

7.3.1   IEEE 802.11

IEEE 802.11 is a standard developed for wireless LAN (WLAN) applications intended to replace conventional wired LANs so that the same applications can run seamlessly with media in 802.3 and 802.5 standards. Nodes in such networks would be mostly laptops and other typical equipment connected to a LAN. The distributed coordination function (DCF) in IEEE 802.11 is the access method used to support asynchronous data transfer on a best-effort basis when the network functions in an ad hoc mode. DCF can also coexist with an infrastructure network.

This is a contention-based protocol based on MACA [8] and MACAW [9] schemes proposed as improvements to the original CSMA scheme developed in Kleinrock and Tobagi [10]. It uses carrier sense multiple access with collision avoidance (CSMA/CA). Collision detection (CD) is not used because a node is unable to listen to the channel for collisions while transmitting. The scheme attempts to avoid the hidden terminal and exposed terminal problems in the original CSMA scheme.

7.3.1.1   Operation

Each node maintains a backoff counter controlling the channel access. Before a node starts data transmission, it senses the wireless medium. If the medium appears to be idle for a specified period of time (distributed interframe space — DIFS), it starts decrementing the backoff counter. If the carrier is detected during this time, the backoff counter is frozen; otherwise, it starts transmission once the backoff counter reaches zero. The sender and receiver exchange short request-to-send (RTS) and clear-to-send (CTS) control frames to establish a session. Data transmission is followed by an acknowledgment (ACK) frame to confirm successful reception. The gaps among RTS, CTS, DATA, and ACK frames are specified by short interframe space (SIFS). Duration of SIFS is relatively shorter than DIFS, thereby giving priority to the ongoing transmission. Contention for the channel access between two nodes, N1 and N2, is illustrated in Figure 7.2. Initially N1 is transmitting frame 1 followed by ACK reception. After waiting for the DIFS period, it starts decrementing the backoff counter in an attempt to transfer another packet. Because the backoff counter of N2 reaches zero first, it captures the medium and transmits a frame while N1 senses the medium is busy. Following the transmission of N2, N1 recaptures the medium for transmission of its second frame.

The DCF adopts a slotted binary exponential backoff mechanism to select the random backoff interval in case of unsuccessful transmission or after the completion of a successful transmission. This random number is drawn from a uniform distribution over the interval [0, CW-1], where CW is the contention window. After an unsuccessful transmission, CW is doubled; once CW reaches a maximum value (CW_{max}), it will remain there. In the case of a successful transmission, the CW value is reset to a minimum value (CW_{min}).

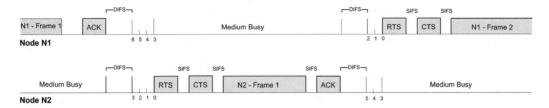

FIGURE 7.2 Contention between nodes N1 and N2 in IEEE 802.11 DCF.

The control frames RTS and CTS, as well as the data frames, include a parameter indicating the expected data transfer duration for the current session, which is used by the other nodes to update their network allocation vector (NAV). NAV is used to maintain a timer at each node, thus avoiding unnecessary transmission attempts before the current transmission is completed. This is termed a *virtual carrier sensing*. During the backoff period and the NAV timer active period, the node will be in idle mode listening to the channel with no transmission attempts.

7.3.1.2   Power-Saving Mode in IEEE 802.11

IEEE 802.11 standard also defines a power saving (PS) mode in which certain nodes can "go to sleep." Under DCF operation, PS nodes "wake up" periodically for a short interval called the Ad Hoc Traffic Indication Map (ATIM) window. It is assumed that hosts are fully connected and synchronized so that the ATIM windows of all PS hosts will start at about the same time. During this window, each node will contend to send a beacon frame. Any successful beacon serves the purpose of synchronizing node clocks and also inhibits other hosts from sending their beacons. After receiving the beacon, an active node can send a direct ATIM frame to a node in PS mode. These transmissions are also contention based and use the same DCF access procedure described earlier. On reception of the ATIM frame, the PS node will reply with an ACK and remain active for the rest of the period. Data transfer will take place after that.

7.3.1.3   Merits, Drawbacks, and Implications for WSNs

Recent work has shown that the energy consumption using IEEE 802.11 MAC protocol is significantly high because the nodes are listening to the channel most of the time. Although the 802.11 standard defines the PS mode, it provides very limited policy about when nodes should go to sleep. PS mode is designed for single-hop networks in which all nodes can hear each other. When used in multihop networks, IEEE 802.11 may have problems in clock synchronization, neighbor discovery, and network partitioning — thereby degrading the performance. Clock synchronization in a multihop WAHN is difficult because there is no centralized control; also, the synchronization packets exchanged among neighbors have variable delays due to unpredictable node mobility and radio interference.

PS mode is typically supported by letting low-power nodes wake up only at specific times. Without precise clocks, a host may not be able to know when other PS hosts will wake up to receive packets. Furthermore, a host may not be aware of a PS host at its neighborhood because a PS host will reduce its transmitting and receiving activities so that it cannot be detected. Such incorrect neighbor information may be detrimental to most routing protocols because the route discovery procedure may incorrectly report that there is no route even when routes exist with some PS hosts in the middle. Tseng et al. [11] proposed three sleep schemes to improve the PS mode in the IEEE 802.11 for its operation in multihop networks.

Requirements for clock synchronization and the suboptimal power saving makes this scheme an improper candidate for medium access in WSNs. Nevertheless, the idea of having a portion of nodes sleeping in the network while others are active may be an applicable concept to WSNs. The presence of redundant nodes in a WSN implies that all the nodes need not be active all the time because other nodes in the neighborhood can perform sensing and communication tasks covering the target area. Therefore, properly chosen redundant nodes can be put into sleep mode to achieve network-wide power savings. Open issues to be explored include the selection of redundant nodes and wake-up and connection reestablishment procedure for these nodes.

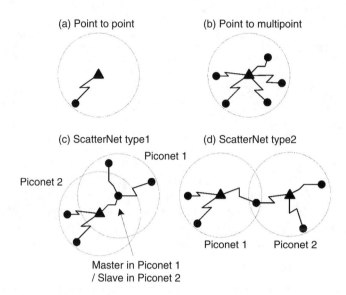

FIGURE 7.3 Piconet configurations in Bluetooth.

7.3.2   Bluetooth

Bluetooth [3, 4] is a short-range wireless networking for electronic consumer devices (mobile phones, pagers, PDAs, etc.). It uses a TDMA and CDMA hybrid scheduling based MAC scheme. The topology is a star network in which several slave nodes are attached to and synchronized with a master node to form a piconet. The number of nodes in a piconet is limited to eight in order to keep a high-capacity link among all the units and to limit the overhead required for addressing. Basic piconet configurations are shown in Figure 7.3(a) and (b). Along with the basic TDMA scheme, Bluetooth uses frequency hopping code division multiple access (FH-CDMA), which uses a large number of pseudorandom hopping sequences. Interpiconet communication is achieved by forming ScatterNets as shown in Figure 7.3(c) and (d). A single node can be a master in one piconet while it is a slave in another. Also, a node can be a slave in two piconets.

7.3.2.1   Operation

The master node determines hopping sequence, provides clock synchronization information for each slave node, and also controls the traffic in the piconet. The master/slave role is only attributed to a unit for the duration of the piconet. When a piconet is cancelled after a certain period of time, the master and slave roles are also cancelled and new piconets will be formed. Any node can become a master or slave. By definition, the unit that establishes the piconet becomes the master. Mechanisms are in place for multiple piconets to interconnect and form a multihop topology.

The time slots are alternately used for master and slave transmissions. The master transmission includes slave address of the unit for which the information is intended. In order to prevent collisions on the channel due to multiple slave transmissions, a polling technique is used: for each slave-to-master slot, the master decides which slave is allowed to transmit. This decision is performed on a per-slot basis: only the slave addressed in the master-to-slave slot directly preceding the slave-to-master slot is allowed to transmit in this slave-to-master slot. If the master has information to send to a specific slave, it is polled implicitly and the slave can return information. If the master has no information to send, it must poll the slave explicitly with a short poll packet. Because the master node schedules the traffic in the uplink and the downlink, intelligent scheduling algorithms that take into account the slave characteristics must be used. The master node control effectively prevents collisions among the participants of the piconet. Independent collocated piconets may interfere when they occasionally use the same hop carrier.

7.3.2.2   Merits, Drawbacks, and Implications for WSNs

Compared to contention-based MAC schemes, TDMA schemes have a natural advantage of energy conservation because the duty cycle of the radio is reduced and there are no contention-introduced overheads or collisions. Nodes can be put to sleep to save energy during the off intervals of the duty cycle, thereby making this an obvious candidate for WSNs.

Use of a TDMA protocol usually requires the nodes to form real communication clusters such as the piconets described here. Nodes in such clusters are restricted to communicate within the cluster, except for the master node and possible gateway nodes. Managing intercluster communication and interference is not an easy task. Moreover, when the number of nodes within a cluster changes, it is not easy for a TDMA protocol to change its frame length and time slot assignment dynamically. Thus, its scalability is normally not as good as that of a contention-based protocol.

In Bluetooth, nodes within a piconet must be synchronized to use the TDMA scheme. Achieving local synchronization within the cluster is not a difficult task. However, network-wide synchronization will be almost impractical, especially in WSNs. Thus, proper mechanisms need to be developed for intercluster communication, perhaps based on contention-based schemes.

7.3.3   Energy-Conserving Medium Access Control (EC-MAC) Protocol for Wireless ATM Networks

This particular MAC protocol is briefly described here because of its significant contribution toward minimizing the power consumption of nodes in wireless and mobile ATM networks. Goals of this access protocol are to conserve battery power; to support multiple traffic classes; and to provide different levels of service quality through bandwidth allocation. Although the IEEE 802.11 and Bluetooth standards address energy efficiency, this was not one of the central design issues in developing these protocols. The EC-MAC protocol [6], on the other hand, was developed with the issue of energy efficiency as a primary design goal.

7.3.3.1   Operation

The EC-MAC protocol is defined for an infrastructure network with a single base station serving mobiles in its coverage area. This definition can be extended to an ad hoc network by allowing the mobiles to elect a coordinator to perform the functions of a base station. Transmission in EC-MAC is organized by the base station into frames and each frame equals the basic unit of wireless data transmission.

The frame structure of EC-MAC protocol is shown in Figure 7.4. At the start of each frame, the base station transmits the frame synchronization message (FSM), which contains synchronization information and the uplink transmission order for the subsequent reservation phase. During the request and update phase, each registered mobile transmits a new connection request according to the transmission order received in the FSM. Collisions are avoided during this phase by having the base station send the explicit order of transmission using the FSM.

New mobiles that have entered the cell coverage area register with the base station during the new-user phase. Collisions during this phase are unavoidable and thus it may be operated using a variant of ALOHA. This phase also provides time for the base station to compute the data transmission schedule. The base station broadcasts a schedule message that contains the slot assignments for the subsequent uplink and downlink data transmissions (see Figure 7.4). Downlink transmission from the base station

FIGURE 7.4   Frame structure in EC-MAC.

to the mobile is scheduled considering the QoS requirements; similarly, the uplink slots are allocated using a suitable scheduling algorithm.

7.3.3.2   Merits, Drawbacks, and Implications for WSNs

Energy consumption is reduced in EC-MAC due to the use of a centralized scheduler, as in Bluetooth. Therefore, collisions over the wireless channel are avoided, thus reducing the number of retransmissions. Additionally, mobile receivers are not required to monitor the transmission channel as a result of communication schedules. The centralized scheduler may also optimize the transmission schedule so that individual mobiles transmit and receive within contiguous transmission slots. This scheme highlights the fact that scheduling algorithms that consider mobile battery power level in addition to packet priority may improve performance for low-power mobiles. Techniques used to minimize the energy consumption and performance of EC-MAC in this regard are discussed in detail in Sivalingam et al. [6].

In contrast to Bluetooth, this scheme allows new mobile nodes to join the cluster without completely disassembling it. In certain WSN applications, the network may consist of a significant portion of mobile nodes among the stationary nodes. Certain stationary nodes may act as sink nodes analogous to the base stations or cluster heads discussed here. When mobile nodes roam around these clusters, a concept similar to EC-MAC can be used to attach new mobile nodes to an existing group or cluster. After such an attachment and schedule update, the mobile nodes can communicate with the cluster head, using the set schedule, at a minimal expense of its energy.

7.3.4   Power-Aware Multiple Access (PAMAS) Protocol

PAMAS (power-aware multiple access) is a contention-based protocol [7] designed for ad hoc networks with energy efficiency as the primary design goal. It modifies the MACA protocol [8] by providing separate channels for RTS/CTS control packets and data packets (out-of-band signaling), thereby avoiding overhearing among neighboring nodes.

7.3.4.1   Operation

In PAMAS, a mobile with a packet to transmit sends an RTS message over the control channel and awaits the CTS reply message from the receiving mobile. If CTS is received, then the node transmits the packet over the data channel. This procedure is shown with nodes N3 and N4 in Figure 7.5. With the start of receipt of the data packet, the receiving mobile transmits a busy tone (BT) over the control channel with more than twice the duration of RTS/CTS packets, thus enabling users tuned to the control channel to know that the data channel is busy. Also, if it hears any other RTS packets (from node N6 in Figure 7.5), it transmits a busy tone.

If an idle node receives an RTS, it will check whether any of its neighbors is transmitting (by sensing the data channel) or receiving (by sensing BT). In either case, it will not reply with CTS (shown with nodes N2 and N5), thus causing the sender of RTS (nodes N1 and N6) to back off using a binary exponential backoff (BEB) scheme. Power conservation is achieved by requiring mobiles that are not able to receive or send packets to turn off the wireless interface. The use of a separate signaling channel allows nodes to determine when and for how long to power off. A mobile should power off when: (1) it has no packets to transmit and a neighbor begins transmitting a packet not destined for it, or (2) it has packets to transmit but at least one neighbor pair is communicating.

Once a node is powered off, two main issues arise. First, the latency due to sleeping is an issue because, if some other node wants to transmit data to a sleeping node, it must wait until this node powers up again. However, it should be noted that, even if the node was awake in this scenario, the sender must wait until the other transmissions are finished. This is a valid argument as long as the node will wake up as soon as the neighboring transmissions and receptions are complete. Therefore, the mechanisms that a node will use to decide exactly when to wake up are crucial.

The second issue is determining the length of sleep duration. It is addressed using a special probe packet. When a node wakes up after some time, it will send out a probe packet over the signaling channel

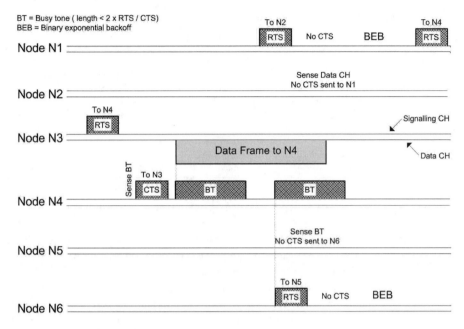

FIGURE 7.5 Operation of PAMAS.

asking any receiving nodes how much time it will take for the current transmission to end. If no collision occurs, the querying node will receive the exact time from the receiving node and will go back to sleep until this time. If the probe packet was destroyed due to a collision, the node will continue a binary search, sending more probe packets. Thus, the use of these probe packets ensures that the node sleeps no longer than necessary, thereby leaving the latency and throughput unchanged.

7.3.4.2 Merits, Drawbacks, and Implications for WSNs

Simulation results have established that this method reduces the power consumption by more than 50% in fully connected networks and at least 10% in highly loaded sparse networks. The essence of this scheme lies in the introduction of the additional signaling channel. This is also a major drawback of the scheme because employing an additional signaling channel requires additional hardware to be built into the nodes. It poses additional challenges, especially in WSNs in which node hardware is highly miniaturized.

Nevertheless, the significance of this protocol is that it can achieve high energy savings without compromising the network throughput and delay. Perhaps the concept of out-of-band signaling used here may be adaptable for intercluster communication in WSNs in which lack of synchronization does not permit using scheduling-based schemes. Such an out-of-band signaling channel can be used to set up data transfer directly between cluster heads or via gateway nodes. This will be remarkably effective for event-driven sensor networks (discussed in Section 7.4.2) in which intercluster communication occurs mostly in cases of a detected event and the regular communication is restricted primarily to nodes within the clusters.

A major contribution of the PAMAS protocol is the power savings achieved without sacrificing network throughput and latency. However, a major drawback observed here is that the power consumption of the nodes during excessive switching between the sleep and wake-up states is not given due attention. With the present WSN hardware designs, power consumption during state switching is significant. At the face of this, the PAMAS method may not perform satisfactorily without appropriate modifications for WSNs. Table 7.2 provides a brief comparison of these four medium access methods for WAHNs.

TABLE 7.2 Comparison of Media Access Protocols for Wireless ad hoc Networks

Protocol	Applications	Features	Implications to WSNs
IEEE 802.11 DCF	Wireless LAN	Optional PS mode	Redundant nodes can be sent to sleep and wake up as need arises
Bluetooth	Wireless networking for personal consumer devices	Piconets; centralized scheduling	Node clustering; local synchronization among nodes in clusters
EC-MAC	Wireless and mobile ATM networks	Scheduling for mobile nodes	Attaching mobile nodes to clusters without disassembling clusters
PAMAS	MAC protocol designed for WAHNs	Out-of-band signaling	Use of out-of-band signaling for intercluster communication

7.4   Design Challenges for Wireless Sensor Networks

As discussed in the first section, nodes in a WSN possess unique characteristics, especially the energy constraints, compact hardware, low transmission ranges, event- or task-based network behavior, and high redundancy. For a WSN, the extension of its lifetime is the most important issue. Therefore, power awareness is prominent in almost every aspect of the operation of WSNs. Currently, the research related to hardware of WSNs is focused on developing ultra low-power sensors, processors, and radio transceivers. Other drives are to reduce the form factor of batteries and improve technologies for power sources to keep nodes alive in active operation for many years. Meanwhile, software and middleware development is focused on minimizing power consumption during network operation. As highlighted in Section 7.2.1, it is extremely critical that the medium access control scheme be power optimal.

Energy consumption of a WSN occurs in three domains: sensing; data processing; and communications; among these, radio communication is the major consumer of energy. As highlighted in Pottie and Kaiser [12], energy for transmitting 1kb over a distance of 100 m is estimated as 3 J. With the same amount of energy, a general purpose processor with 100 MIPS/W power could execute 3 million instructions. The sensing circuitry consumes less power than the processor board in a typical WSN platform such as MICA [13, 14]. However, the radio consumes two to three times the power of the processor during packet transmission. Power consumption of the radio during listening to the channel for reception is also higher than the processor at full operation, but relatively lower than the transmitting power. The MICA sensor network platform defines four modes of operation, and Table 7.3 shows the typical current draw and power consumption of each node.

Thus, it is clear that the research focus should be on optimizing the medium access method in order to extend the lifetime of the network. In addition to energy conservation, the ability of the MAC scheme to adapt to network size, node density, and topology is also important. To be used in sensor networks aimed for dynamic applications, a MAC scheme should be highly scalable. Other important attributes include fairness, latency, throughput, and bandwidth utilization. However, these issues are considered secondary compared to energy considerations because they determine the entire lifetime of the network.

TABLE 7.3   Modes of Operation in MICA

Mode	Typical Current Draw	Power Consumption
Transmit (peak power)	32 mA	95 mW
Receive	18 mA	55 mW
Idle/sense	8 mA	25 mW
Sleep	20 μA	60 μW

Source: Crossbow Inc., Data sheet for MICA2 wireless measurement system, 2003.

Similar to the schemes described in Section 7.3, MAC schemes for sensor networks can be fundamentally categorized into *contention-based* or *scheduling-based* schemes. The inherent advantages of contention-based schemes in the context of WSNs include:

- No synchronization requirements
- No central scheduler required
- More robust to network dynamics
- No clustering necessary
- More suitable for event-driven WSNs

However, in terms of energy savings, contention-based schemes are not very attractive. Several sources of energy wastage in contention-based schemes during communication [15] can be identified:

- *Collision*. Usually data gathered by a node are exchanged with others using the radio. Two nodes may transfer data to each other at the same time or several nodes transfer data to the same node at the same time. When a transmitted packet is corrupted, it must be discarded and, thus, the follow-on retransmissions increase energy consumption. Collision increases latency as well.
- *Overhearing*. When a node picks up packets destined to other nodes, overhearing occurs. In an ad-hoc fashion, a transmission from one node to another is potentially overheard by all the neighbors of the transmitting node; thus, all of these nodes consume power even though the packet transmission was not directed to them.
- *Control packet overhead*. Sending and receiving control packets such as routing updates consumes energy and effectively reduces the network bandwidth for data packets.
- *Idle listening*. Nodes must listen to the channel often in order to receive possible traffic that is not sent. This is especially true in many sensor network applications because, if nothing is sensed, nodes are in idle mode for most of the time. Actual measurements have shown that idle listening consumes 50 to 100% of the energy required for receiving in such networks.

Scheduling-based schemes attempt to determine network connectivity first (i.e., discover the neighbors of each node) and assign collision-free links to each node. Links may be assigned as time slots (TDMA), frequency bands (FDMA), or spread spectrum codes (CDMA). However, the miniature hardware design of nodes in a WSN may not permit employing complex radio transceivers required for FDMA or CDMA systems. Thus, TDMA schemes are preferred as scheduling methods for WSNs. Inherently, TDMA schemes have a distinct advantage over the other methods. Except for the transmission, receiving and sensing durations, nodes can be put to sleep in order to achieve the highest amount of energy savings possible.

Nevertheless, the task of assignment of channels (i.e., TDMA slots, frequency bands, or spread spectrum codes) to links between neighbors so that packets do not collide is difficult. To ease the assignment, often a hierarchical structure is formed in the network to localize groups of nodes and make the task of channel assignment more manageable. This requires formation of node clusters and elect leaders for each cluster.

For TDMA schemes, time synchronization is a crucial factor, also. In contrast to generic WAHNs, maintaining perfect synchronization over the whole network is almost impossible, mainly because of the wide range of deployment, lower transmission ranges, and less control packet transmissions permitted due to energy constraints. Under a hierarchical clustering scheme, synchronization within each cluster can be maintained, but intercluster communication poses problems because of lack of synchronization.

7.4.1 Why Existing Methods for Wireless Ad Hoc Networks Cannot Be Used

The main goals of WAHNs are to provide a high throughput and low delay at a high bandwidth. In such networks, all nodes are engaged in the same type of activity and each user deserves equal opportunity in accessing the media. Thus, per-node fairness is an important issue. Network lifetime is not considered significant because the energy sources can be recharged or replaced, although power-saving schemes are recommended.

In contrast, for a WSN, extending the network lifetime is one of the priorities. To this end, it is necessary to conserve energy at each node during network operation. Toward achieving this objective, one must be ready to compromise network throughput and latency to a certain extent. Moreover, based on the type and operation of the WSN, as described in Section 7.4.2, throughput and latency requirements will depend on the application.

Also in contrast to WAHNs, nodes in WSNs are highly redundant; thus, some nodes can afford to be in sleep mode until a need arises, while others are active. Certain nodes acting as cluster heads, gateways, or sink nodes would accumulate, process, and relay larger quantities of data than ordinary, leaf-level sensor nodes. Additionally, from time to time during certain applications, the importance of data sensed by a node may vary in its importance or relevance to the current network objective. These issues call for maintaining distinct node priorities in WSNs in contrast to per-node fairness desired in WAHNs. Thus, employing MAC schemes that are developed for WAHNs would not be satisfactory in sensor networks without proper modifications.

In summary, novel MAC protocols are needed for WSNs because:

- Extending network lifetime is the primary goal in WSNs.
- Throughput and delay performance become secondary goals.
- QoS requirements may vary from time to time (e.g., in an event-driven WSN).
- Per-node fairness is not desired; instead, distinct node priorities may need to be considered for resource allocation.

7.4.2 Communication and Application Types in Sensor Networks

This section attempts to categorize the application-led behavior and possible communication types in WSNs. This is a general categorization that may help to identify relevant issues in designing an optimum medium access scheme for WSNs. Depending on application characteristics, sensor networks will behave in one of the following ways. It is possible that the same network may adopt a different role due to changes in the system objectives or firing of certain events in the observation field.

- *Centralized* data gathering and decision making. These networks are hierarchically organized, thus easier to set up and manage. At the top of the hierarchy, one or more root (sink) nodes collect all data from leaf nodes. Local processing may be performed at the sensor nodes at the bottom of the hierarchy, but the root node is responsible for gathering final data and, for the most part, governing the operation of the whole network.
- *Distributed* data gathering and decision making. Tasks in these networks are highly distributed and it is difficult to identify a particular network architecture or a hierarchy. An example would be a set of sensor nodes dropped in a harsh environment with no central control. Individual nodes must perform whatever sensing operations they can, discover their own neighbors and perhaps collectively make decisions on discovered events, and relay the decisions to the outside world via any relay nodes in reach. Instead, they would control certain actuators within their reach to perform certain reactive actions individually or by temporarily appointed leader nodes.

In both data gathering schemes, the following types of communication can occur:

- Unicast messages:
  - *Local.* When a real-world event in the network occurs, nodes are expected to perform some in-network processing. This will generally involve local messages being exchanged between neighbors. In a cluster-based scheme, this will include the messages exchanged between two nodes in the cluster or between a member node and the cluster head.

- *Multihop.* In centralized data gathering applications when a node requires sending data to the sink node (node-to-sink reporting), the sink node will not be in its direct reach most of the time. Thus, it will pass the message with the intended recipient address over multiple hops. This needs a proper addressing scheme as well as an efficient routing scheme.
- Multicast messages:
  - *Local.* These are messages originating from a node intended for several neighbors within its direct transmission range. For a clustering-based scheme, this is limited to multicast within the cluster.
  - *Multihop.* An example is a situation in which a sink node or a root node requires passing a control message to a set of nodes. In a clustering-based scheme, this may be communication from a root node to a set of cluster heads, or from one cluster head to several others. All such multihop messages need a proper addressing scheme as well as an efficient routing scheme, as mentioned earlier.
- Broadcast messages:
  - *Local.* Messages will be broadcast by a node to all the neighbors within its reach. Such messages will include anything of local importance to the neighborhood. In a clustering-based scheme, these will include messages broadcast among all nodes in a group.
  - *Multihop.* These are the messages that will impose the heaviest communication burden on the network. For instance, in a monitoring and surveillance application, a node may observe an alarming condition and may need to alert all others in the network. Unrestricted flooding may not be appropriate for such a situation, but a combination of multihop–multicast with local broadcast may be used.

Based on the application, optimal communication strategy for a sensor network would also be different. Two major categories of sensor networks dictated by their applications [16] have been identified:

- *Event driven sensor networks.* In an event-driven sensor network, sensor nodes do not send data (and are most likely asleep) until a certain event occurs. For example, in a fire-monitoring application, until a rise in temperature or smoke is detected, no data need be sent. In this way, node energy can be maximally saved. When an event occurs, how quickly the event can be reported to a central station, or how quickly other neighboring nodes can be alerted, become important issues. The main difficulty in an event-driven sensor network is to wake up the entire network, or at least the nodes along a path to the base station, when an event occurs. Moreover, the traffic pattern of the network may drastically change in case of an event.
- *Continuous monitoring sensor networks.* In a continuous monitoring sensor network, data are sampled and transmitted at regular intervals. For example, an ambient temperature monitoring station can take periodic readings and send it to a central monitoring station only at specific intervals. In these types of networks, the traffic patterns are more stationary and the routing tables (if any) remain unchanged most of the time. Scheduling-based MAC schemes can be used effectively in these networks for maximum energy savings.

The behavior and communication types identified in this section need to be considered for the optimal performance of energy-aware MAC schemes. Next, several MAC schemes proposed for WSNs will be discussed, along with their applicability to these different types of networks.

7.5 Medium Access Protocols for Wireless Sensor Networks

Recently, several authors have suggested energy-aware medium access schemes for WSNs, a number of which are modifications of existing protocols for WAHNs. This is still a growing area of research calling for attention to several open issues yet to be addressed. This section discusses four such recently proposed schemes, with their merits and drawbacks, in the context of WSNs. These include sensor MAC (SMAC) [15]; self-organizing MAC for sensor networks (SMACS) [17]; traffic adaptive medium access protocol

(TRAMA) [18]; and power-efficient and delay-aware medium access protocol for sensor networks (PEDAMACS) [19].

7.5.1   Sensor MAC (SMAC)

The main objective of SMAC [15] is to conserve energy in sensor networks; it takes into consideration that fairness and latency are less critical issues compared to energy savings. Thus, this scheme compromises fairness and latency to a certain degree. In order to save energy, SMAC establishes a low duty cycle operation in nodes. It reduces idle listening by periodically putting nodes into sleep in which the radio transceiver is completely turned off. As discussed in Section 7.2.1, a high bandwidth utilization is a goal in generic WAHNs, compelling nodes to operate in fully active mode all the time. In SMAC, the low duty cycle mode is the default operation of all nodes in the network. Nodes only become more active by changing the duty cycle when heavy traffic is present in the network, or once an event occurs in case of an event-driven WSN.

7.5.1.1   Operation

During the design of SMAC, the following assumptions have been considered:

- Short-range multihop communications will take place among a large number of nodes.
- Most communications will be between nodes as peers, rather than to a single base station.
- In-network data processing is used to reduce traffic.
- Collaborative signal processing is used to reduce traffic and improve sensing quality.
- Applications will have long idle periods and can tolerate some latency.
- Network lifetime is critical for the application.

All nodes in the network will be following a sleep-and-listen cycle called a frame, as shown in Figure 7.6. The duration of the listen period is normally fixed and the sleep interval may be changed according to application requirements, changing the duty cycle.

The same RTS/CTS/DATA/ACK procedure as that in IEEE 802.11 is adopted here for unicast packets in order to ensure collision avoidance and to avoid hidden terminal problem. Broadcast packets are sent without using RTS/CTS; NAV timer update information is included in all four types of packets. Thus, this scheme uses virtual and physical carrier sensing. After a successful exchange of RTS and CTS, the sender will start transmission and will extend it into the sleeping duration as well, if required. The nodes do not follow their sleep schedules until they finish the transmissions, thus increasing the performance.

7.5.1.2   Coordinated Sleeping

Although a node can freely choose its own active/sleep schedules in SMAC, it attempts to reduce the overhead by synchronizing schedules of neighboring nodes together. Nodes exchange their schedules by periodically broadcasting a SYNC packet to their immediate neighbors at the beginning of each listen interval. A set of nodes synchronized together will form a virtual cluster. Because the whole network cannot be synchronized together, neighboring nodes are allowed to have different schedules. However, neighboring nodes are free to talk to each other, no matter to which listen schedules they adhere. A considerable portion of the nodes will belong to more than one virtual cluster, enabling intercluster communication. Thus, this scheme is claimed to be adaptive to topology changes.

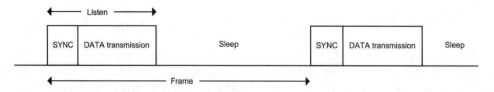

FIGURE 7.6 Periodic listen-and-sleep schedule in SMAC.

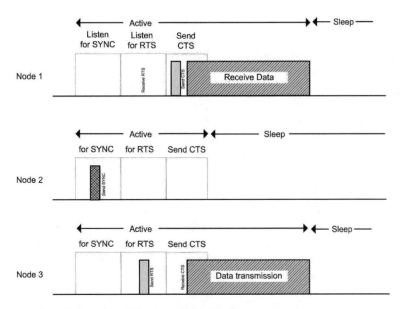

FIGURE 7.7 Timing schedules among different nodes in SMAC.

When the network is first deployed, each node tries to retrieve a sleep schedule from a neighbor first. In case of failure, it adopts one of its own and also tries to announce it to the neighbors by broadcasting a SYNC packet. Broadcasting SYNC packets must also follow the normal carrier sense and random backoff procedure. If a node receives a different schedule after it announces its own schedule, it must adopt one of the following:

- If the node detects no other neighbors, it can discard the current schedule and adopt the new.
- If it has one or more neighbors and is already a part of an existing virtual cluster, it can adopt both schedules by waking up at the listen intervals of both.

The active interval of a node is divided into three parts for SYNC, RTS, and CTS as shown in Figure 7.7. If CTS is received, data transmission will be immediately followed. Here nodes 1 and 3 are synchronized to the schedule of node 2 by receiving its SYNC packet, thus falling into the same virtual cluster. Node 3 initiates an RTS/CTS exchange with node 1 followed by a data transmission. While node 2 follows its normal sleep schedule, nodes 1 and 3 stay active until the completion of the data transfer, altering their usual schedule.

7.5.1.3 Neighbor Discovery in SMAC

When a new node powers on, it listens to the channel in anticipation of a SYNC packet. However, it is possible that a new node fails to discover an existing neighbor because of collisions or delays in sending SYNC packets by neighbor due to busy medium. To prevent a case in which two neighbors cannot find each other when they follow completely different schedules, SMAC protocol employs a simple periodic neighbor discovery procedure by requiring each node to listen periodically to the channel for the whole synchronization period. The frequency can be varied depending on the network conditions, etc.

7.5.1.4 Synchronization

Clock drift on each node can cause errors in the coordination of schedules among neighboring nodes. To minimize this problem, it uses relative timestamps in SYNC packets. Also, the listen period is made significantly longer than possible clock drift. Although this technique can tolerate relatively larger clock drifts, neighboring nodes are still required to update each other periodically with their schedules to prevent possible errors. Using experiments, authors claim that the clock drift between two nodes does not exceed 0.2 ms per second [15]; however, these figures may not be valid for certain applications of WSNs.

7.5.1.5   Adaptive Listening

An adaptive listening strategy that enables each node to adjust its schedule according to the network traffic is also used in SMAC to minimize latency. When a sensing event occurs, it is desirable that the sensing data can be passed through the network without much delay. When each node strictly follows its sleep schedule, potential delays are possible on each hop in the multihop path. The technique is that, if a node overhears its neighbor's RTS/CTS transmission during a listen period, it will receive the estimated length of that data transmission before going to sleep according to its normal schedule. However, the node will wake up for a short period of time at the end of that transmission to check whether it is the next hop in this multihop message. If so, the neighbor will immediately pass the data to it after RTS/CTS exchange, thus avoiding the neighbor's waiting for the next scheduled listen time of this node and minimizing latency.

7.5.1.6   Merits and Drawbacks

Compared to other schemes designed for the mobile ad hoc networks explained in Section 7.3, SMAC is designed particularly for use in wireless sensor networks. It attempts to combine the advantages of TDMA scheduling for power saving by periodically requiring sensing nodes to go to sleep. The sleeping patterns are coordinated in order to minimize the latency, as discussed before. Nevertheless, a solution with a fixed duty cycle does not give the optimal performance.

Authors claim that this scheme forms a flat topology and intercluster problems are absent; however, this may not be true in cases in which the application requires real clusters to be formed, at least temporarily. In such a case, the communication patterns will depend on the cluster formation and that these real clusters and the virtual clusters formed will coincide is not guaranteed. The adaptability of this scheme to such a situation should be investigated.

A significant portion of nodes will belong to two or more virtual clusters under this scheme. The energy consumption of such nodes would be higher compared to nodes within a single virtual cluster. Hence, the portion of such nodes and its effect on performance should be analyzed under real application scenarios. Also, the performance of this MAC scheme should be studied along with different routing schemes in order to assess its performance of intercluster communication, especially for multihop unicast and multicast messages. Data routing across virtual clusters needs to be studied further for its latency and throughput.

In WSN applications, it is possible for certain nodes to be exhausted with power and new nodes to be added. Performance of SMAC during times when a significant portion of nodes is discarded or added, or in cases with a higher portion of mobile nodes, should be studied. Another instance to be observed is what happens if the coordinated sleep schedules of two neighboring clusters are completely opposite. In such cases, it is not clear whether the bordering nodes could adopt both schedules.

7.5.2   Self-Organizing MAC for Sensor Networks (SMACS) and Eavesdrop and Register (EAR) Algorithms

Self-organizing MAC for sensor networks (SMACS) is designed for network startup and link layer organization in a static WSN [17]; it is one of the earliest attempts to develop MAC protocols for sensor networks. In this scheme, each node maintains a TDMA frame in which the node schedules different time slots to communicate with its known neighbors. During each time slot, it only talks to one neighbor. To avoid interference between adjacent links, the protocol uses different frequency channels (FDMA) or spread spectrum codes (CDMA). Although the frame structure is similar to a typical TDMA frame, it does not prevent two interfering nodes from accessing the medium at the same time. The actual multiple access is accomplished by FDMA or CDMA.

7.5.2.1   Operation

For the correct operation of the SMACS protocol, the following assumptions are made.

1. Nodes are able to tune the carrier frequency to different bands. It is assumed that the number of available bands is relatively large.

2. Nodes are randomly deployed. After deployment, each node wakes up at some random time according to a certain distribution.
3. The network is assumed to consist primarily of stationary nodes, with few mobile nodes.

Each node assigns links to its neighbors immediately after they are discovered. When all nodes hear all their neighbors, they have formed a connected, multihop network. Because each node is only partially aware of the radio connectivity in its vicinity, it is possible that collisions can occur if a simple TDMA scheme is used alone. To avoid this, frequency bands chosen at random from a large pool are assigned for each slot.

Length of the frame T_{frame} is fixed for all nodes in the network; however, these frames need not be synchronized and the time slots assigned inside the frame need not be aligned. This is possible because different frequency band or CDMA codes are used for communication during each slot. Such an ability to assign nonsynchronous slots is the key property that enables nodes to form links on the fly. This is illustrated in Figure 7.8, in which nodes A, B, C, and D are in the same neighborhood after deployment and they wake up at times T_1 to T_4, respectively.

Nodes A and B discover each other first and establish their own schedules for transmission and reception. Nodes C and D wake up at later times, discover each other, and establish their own schedules. However, note that all schedules are not aligned and also that the transmission slots of pair A/B and pair C/D overlap. This is made possible by using distinct frequencies f_x and f_y. After a schedule is established, a node will turn on its transceiver ahead of appropriate slots to communicate with others. Similarly, it will turn off the radio when no communication is scheduled, thereby enabling significant energy savings. In most WSN applications, mobile nodes will be present among other stationary nodes. In order to attach these mobile nodes in an energy conserving manner to the already formed network using SMACS, the eavesdrop and register algorithm (EAR) is introduced and discussed in the following section.

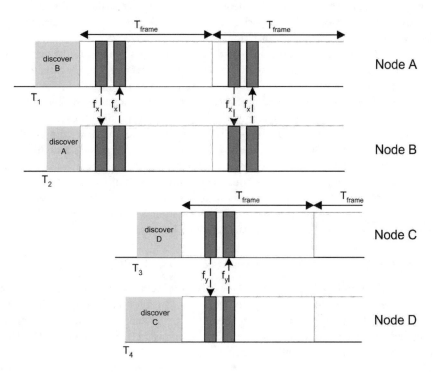

FIGURE 7.8 Nonsynchronous scheduled communication in SMACS.

7.5.2.2   EAR Algorithm

The EAR algorithm enables seamless interconnection of mobile nodes in the field of stationary wireless nodes. This protocol performs the mobility management of the network allowing mobile nodes to listen to the communication from the stationary nodes and establish connectivity with them. Because of energy limitations, the communication channels between the mobile and stationary sensors in the network must be established using as few messages as possible. This is accomplished by allowing the mobile node to decide when to invite the stationary node to establish a connection as well as when to drop a connection. In this manner, mobile nodes assume full control of the connection process to avoid the unnecessary use of power associated with lost messages.

According to the preceding third assumption, only a few stationary sensors will be within the reach of a mobile sensor at any given time. During some predetermined slot in the frame of each stationary node, it transmits an invitation message to the surrounding neighborhood with the intent of inviting new nodes to join the local network. Stationary nodes do not necessarily require a response to this message, but a mobile node with the intention of joining the network will be eavesdropping on such messages. These pilot messages will trigger the EAR algorithm in mobile nodes.

Each mobile node will maintain a list of neighbors according to the invitation messages received. It compares parameters, such as the received SNR, node ID, transmitted power, etc., and decides which node to connect. When the energy saving requirements are stringent, the decision will aim solely for minimal power connectivity. Accordingly, the mobile nodes will initiate a connection with a stationary node. Stationary nodes will also maintain a simple list of mobile nodes that have formed connections and remove the entries when the link is broken.

7.5.2.3   Merits and Drawbacks

Merits of this scheme include the ability to form links with any neighbor on the fly, with no restrictions on synchronization, which largely reduces the latency. Another advantage is that this scheme is applicable to WSNs with neither physical nor virtual clustering. This will allow the MAC scheme to function independently of any application-based clustering requirements of WSN.

However, a significant waste of resources is a trade-off to low latency. The main drawback is that the time slots are wasted if a node does not have data to send to the intended receiver. Also, the frames of a large number of nodes will mostly be vacant if the nodes are sparsely distributed in certain areas. Defining a smaller T_{frame} value will not be permitted in this case because some areas of the network may have a higher density of nodes. SMACS does not attempt to utilize these vacant time slots in order to maintain simplicity; rather, this protocol uses FDMA or CDMA and thus unnecessarily complicates the node hardware design. Bandwidth utilization would also be lower for the same reasons. Another major drawback is that the energy waste during the switching between sleep and active states is not considered. Because of assigning time slots on the fly without any synchronization, the nodes must switch between active and sleep states many times. This will drain the energy sources of nodes unnecessarily.

Apart from these drawbacks, this EAR protocol should be studied further in order to develop effective ways to manage WSNs with mobile as well as stationary nodes.

7.5.3   Traffic Adaptive Medium Access Protocol (TRAMA)

TRAMA is a recently introduced MAC protocol for energy-efficient and collision-free channel access in WSNs [18]. It uses traffic-based information to decide on schedules for individual nodes and thus is adaptive to network traffic. It is claimed that, because of this adaptability, it can deliver adequate performance and energy efficiency in both network types discussed in Section 7.4.2. TRAMA provides support for unicast, broadcast, and multicast traffic.

7.5.3.1   Operation

TRAMA assumes a single, time-slotted channel for data and signaling transmissions. The time schedule of each node is organized in two major sections, as shown in Figure 7.9. One consists of a collection of

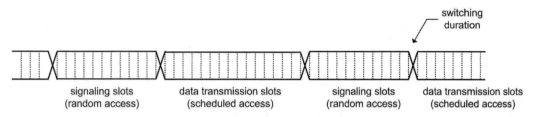

FIGURE 7.9 Time slot organization in TRAMA.

signaling slots using random access and the other of data transmission slots using schedules access. The duty cycle of switching between these states could be adjusted according to the application requirements and also according to the different network types described in Section 7.4.2. For stationary networks, the random access periods occur less frequently and vice versa for highly dynamic networks. Cycle duration is usually of the order of tens of milliseconds, making this scheme less prone to even significant clock drifts around 1 msec, which are highly unlikely in typical networks. Thus, the scheme assumes that adequate synchronization can be achieved using one of the synchronization schemes suggested for WSNs.

Communication in TRAMA consists of three major components: neighbor protocol (NP); the adaptive election algorithm (AEA); and the schedule exchange protocol (SEP). NP is used to exchange one-hop neighbor information among neighbors and to gather two-hop topology information for each node in the network. This is performed by exchanging small packets among neighbors during the random access period. Nodes always start in random access mode with NP. Also, the synchronization is performed during the random access period and the node should be in active state (transmit, receive, or listen) during this interval.

During the random access period, NP exchanges short signaling packets that include the information about connected neighbors of the sender; the goal is to provide the two-hop neighbor information to each node. Note that these include incremental information to keep the packet length small, i.e., it contains the node IDs of newly added neighbors and disconnected neighbors. These short packets are also used to maintain connectivity between neighbors.

During scheduled access, AEA selects transmitters and receivers so that collision-free transmission is achieved. AEA is based on the neighborhood-aware collision resolution protocol (NCR) proposed in Bao and Garcia–Luna–Aceves [20]. This technique claims to avoid data packet collisions among neighbors due to hidden terminals. AEA uses traffic-based information exchanged among nodes during SEP to make efficient use of the channel avoiding idle slots. The same traffic information is also used in AEA to perform receiver selection. During these selections, the node priorities in the network and two-hop neighbor information exchanged during NP are considered. By selecting the transmitter and receiver for each time slot, AEA enables nodes to switch into sleep mode whenever possible, thus achieving maximum energy savings.

SEP is used to exchange traffic schedules among neighbors during scheduled access mode. These schedules contain the set of receivers for the traffic currently originating at the node and its scheduled transmission slots. This information is periodically broadcast to the node's one-hop neighbors during scheduled access. Each node computes a schedule interval depending on how often it needs to transmit data according to its current application requirements. Following this, the node selects the highest priority slots it can acquire according to AEP; an example is illustrated in Figure 7.10. This information (schedule) is transmitted to its neighbors typically during the last slot of its schedule. Also, after emptying the current data buffers, it announces the release of its vacant time slots so that the other nodes can acquire them.

State of a node at a particular time slot is determined based on its two-hop neighborhood information and the schedules announced by its one-hop neighbors. Three possible states of a node are: transmit, receive, and sleep. At a given time slot, a node is in transmit state if it has the highest priority among its contending set and also if the node has data to send. A node is in the receive state when it is the intended

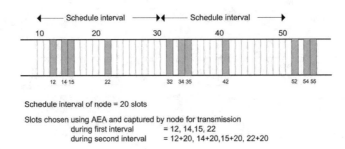

FIGURE 7.10 Example schedule of a node in TRAMA.

receiver of current transmitter. If neither of these cases occurs, the node will be switched off to the sleep state in order to save energy. In other words, if a node is not the currently selected transmitter by AEP, it will consult the schedule sent by the current transmitter. If the transmitter does not have traffic destined for this node in the current slot, it can go to sleep. Under this scheme, a sleeping node is required to wake up at the schedule announcement slot (usually the last transmission slot in each schedule interval) to update itself on possible schedule changes. For this purpose, it should always be aware of the schedule of each of its one-hop neighbors.

7.5.3.2 Merits and Drawbacks

The authors provide extensive simulations to compare TRAMA with SMAC and several other comparable MAC schemes [18]. It is shown that the scheduled-based medium access protocol based on neighbor-hood-aware collision resolution protocol (NCR) achieves better data delivery than the contention-based protocols such as IEEE802.11, CSMA, and S-MAC. The main reason highlighted for the improvement in delivery is that the freedom from collision is guaranteed at all times during data transmission.

It is also shown that the scheduled-based medium access protocols incur higher average queuing delays. The average queuing delay for TRAMA is relatively large due to overhead involved in scheduling. Within every schedule interval, a transmission slot is used for announcing schedules in TRAMA. This decreases the effective channel access probability for data transmission and is not favorable for continuous data gathering type WSNs described in Section 7.4.2 because the traffic is homogenous across the network and all the nodes periodically generate traffic.

Simulation results also show that TRAMA exhibits high throughput compared to SMAC and IEEE 802.DCF because it avoids collisions due to hidden terminals using NCR protocol [20]. Energy savings of TRAMA depend mainly on the traffic pattern of the network as compared to duty cycle-dependent energy savings in SMAC. In TRAMA, the random access period duration plays a significant role in energy consumption. Significant features in TRAMA are the time slot reuse; using neighborhood information for collision avoidance; and use of a hybrid scheme of random and scheduled access for optimal performance.

7.5.4 Power-Efficient and Delay-Aware Medium-Access Protocol for Sensor Networks (PEDAMACS)

PEDAMACS [19] medium access protocol combines the characteristics of cellular networks with those of type 2 sensor networks for the continuous data gathering applications described in Section 7.4.2. It assumes that a single access point (AP) exists in the network and all nodes communicate with this AP. Also, it assumes that AP has no energy constraints and is capable of transmitting at higher power levels when needed so that it can reach any node in the network in a single hop. In contrast, the sensor nodes have limited transmission power and will reach the AP using multiple hops. Although it may not be possible always, in certain applications it may be possible to include a few nodes with higher energy resources to act as APs of each node cluster. The extra effort required may be compensated with optimal power savings in low-power sensor nodes.

7.5.4.1   Operation

The algorithm consists of three major phases: topology learning phase; topology collection phase; and scheduling phase. During the topology learning phase, each node identifies its interferers, neighbors, and parent node. This phase begins with a topology learning packet transmitted by AP over the longest range (highest power) in one hop to all sensors. This packet includes the current time so that each node updates its time and synchronizes with the other. Also, it includes the next anticipated incoming packet time so that every node will stop transmitting and listen for the next broadcast message of AP at this future time, as illustrated in Figure 7.11(a).

Following this, AP floods the network with a tree construction packet over a medium range (medium power). This packet contains a hop count field to avoid any retransmission loops and to facilitate choosing the parent node in the tree as shown in Figure 7.11(b). At the end of this phase, each sensor node decides the parent node to be the one with the smallest number of hops to AP, and the neighbors and interferers as the nodes with the received signal level above and below some interfering threshold, respectively. Because no prior topology information is available during this phase, the authors suggest a simple CSMA scheme with a random delay before carrier sensing.

The topology collection phase starts next with the AP transmitting a topology collection packet (with the same format as that shown in Figure 7.11(a) over the longest range (highest power). The transmission time is announced in the incoming packet time field of the topology learning packet earlier. This packet also contains current time and next incoming packet time. Following this, each node transmits its topology packet containing its parent, neighbor, and interferer information to AP as shown in Figure 7.11(c). Here again, the CSMA scheme with some random delay before the transmission is used.

During the scheduling phase, each node is explicitly scheduled by AP based on the complete topology information obtained during the previous topology collection phase. The scheduling frame is divided into time slots. At the beginning of this phase, AP performs the scheduling of the sensor nodes in the network and announces the schedule of how all the traffic will be carried during the scheduling frame by broadcasting a schedule packet over the longest range. The schedule packet includes the transmitter information corresponding to each time slot in addition to current time and next incoming packet time fields as shown in Figure 7.11(d). At the beginning of the scheduling frame, each node samples the sensor and generates one packet, which is then carried to AP according to the schedule.

Header	Current time	Next packet transmission time	CRC

(a) Topology learning and topology collection packet from AP

Header	Number of hops	Parent transmission node ID	CRC

(b) Tree construction packet from AP

Header	Node ID	Node level	Parent ID	No. of neighbors	Neighbor IDs	No. of interferers	Interferer IDs	CRC

(c) Topology packet from nodes

Header	Slot seq. No.	Number of nodes scheduled for current slot	Scheduled node IDs	. . . . . .	CRC

(d) Schedule coordination packet form AP

FIGURE 7.11 Packet formats in PEDAMACS.

7.5.4.2   Merits and Drawbacks

Although in certain specific applications such a scheme may be able to be used for sensor networks, characteristics of this scheme are not preferred for sensor networks in general. This is mainly due to the use of a central access point that can reach all sensor nodes using high transmitting power. This assumption is not realistic in most of the sensor network applications in which the nodes are distributed over large areas or in indoor environments. Authors argue that in such cases, several APs can be used, but fail to provide details on how to establish proper coordination and synchronization among all such APs. Furthermore, it has yet to be analyzed how the large overhead associated with such a scheme affects the network performance. This overhead also makes this scheme unusable for event-driven sensor systems or dynamic systems with frequent addition and removal of nodes from the network. Authors suggest using a CSMA scheme with implicit ACKs for the topology collection phase, but this may not be an appropriate solution to avoid the huge number of possible collisions.

7.5.5   Comparison

Table 7.4 summarizes and compares the previously discussed four MAC schemes for WSNs.

7.6   Open Issues

Having discussed application and communication categories of WSNs and compared several MAC schemes for WSNs, the open research issues yet to be addressed will be discussed in this section. Designing optimal, energy-aware MAC schemes for WSNs is still an open and fast growing research area. In this context, the following issues are highlighted for consideration in future research.

TABLE 7.4   Comparison of MAC Schemes for Sensor Networks

	SMAC	SMACS/EAR	TRAMA	PEDAMACS
Features	TDMA scheduling Coordinated sleeping schedules among neighbors Adaptive listening Virtual clustering	Hybrid TDMA/FDMA scheduling Mobile node attachment	Random access (CSMA) for neighbor discovery Scheduled access (TDMA) for data transmission	Access point (AP) with high-power transmitter Centralized TDMA scheduling by AP node Hierarchical organization
Applications	WSNs with more stationary nodes	Low traffic WSN with strict latency requirements	Event-driven WSNs	Centralized data gathering WSNs
Merits	Reduced latency for multihop messages Simple hardware for TDMA	Low latency Ability to create links on the fly No clustering requirements No synchronization requirements	TDMA slot reuse No collisions due to hidden nodes Traffic adaptable	Higher energy savings in centralized WSNs
Drawbacks	Synchronization required Virtual clusters may not coincide with physical clusters	Complex hardware for FDMA or CDMA Waste of time slots Low bandwidth utilization Frequent switching can cause heavy energy losses	Synchronization required Low bandwidth utilization in periodic data gathering WSNs	Centralized control necessary AP node requires high power High overhead for scheduling

Adaptability to network objectives. How much sensor node energy to spend on a particular task entrusted on a WSN depends on how critical current application objectives are. As explained in Section 7.2 and Section 7.4.2, the same network used for a low-frequency continuous monitoring application may be employed for mission-critical tracking or emergency threat alert in the next instance. In such a scenario, less critical goals of a sensor network become highly critical and the energy saving requirements become secondary as compared to latency and throughput. A challenging and open issue is to develop medium access schemes for WSNs that have changing missions. SMAC [15] and TRAMA [18] attempt to achieve this to a certain extent; nevertheless, more work must be done in this area.

Optimal schemes depending on WSN type. Certain applications such as habitat monitoring may have stationary traffic patterns mostly over the total lifespan of the WSN employed. For these types of applications, achieving energy savings to extend the network lifetime remains the primary objective throughout the monitoring period. Medium access schemes can be optimized for energy efficiency in WSNs used for such applications. TDMA scheduling-based schemes similar to SMAC [15] may be the ideal candidate for these applications. However, the synchronization requirements and virtual clustering need to be reconsidered in this respect.

Cross-layer design. Conventional WAHNs have neatly defined protocol stacks with independent operation of each layer. For example, medium access scheme would function independent of the node connectivity, routing requirements, and application context. However, the primary goal of energy saving is tightly coupled with all these factors in a WSN and thus medium access cannot be considered alone for optimal savings. It is increasingly clear that power efficiency cannot be addressed completely at a single layer in the networking stack [21, 22]. It will often be necessary to use parameters propagated from upper layers to adapt the medium access protocol, especially in situations in which network objectives change considerably from one time to another. Another issue that must be effectively coupled with medium access is the data aggregation. No significant research efforts have been observed so far in this regard and the next generation medium access schemes for WSNs beyond SMAC and TRAMA should take these aspects into thorough consideration.

Effects of time synchronization. It is observed that higher energy savings are mostly obtained using TDMA based-scheduling schemes in WSNs. Inherently, these schemes require time synchronization of participating nodes in a single schedule. Synchronization errors always tend to degrade the end-to-end throughput performance of the network. As highlighted in Section 7.3.2.2 and Section 7.4, it may be impractical to achieve network-wide synchronization in WSNs. As an alternative, it is better to have globally asynchronous and locally synchronous architectures in which local node clusters maintain synchronization for TDMA schemes aiming for maximum energy savings, while intercluster communication is mostly contention based and asynchronous. Coupling such schemes for optimal energy saving medium access has yet to be explored.

Cluster-based hierarchy. Most WSN applications may require hierarchical architectures. This favors clustering-based systems, which is assumed in most of the research on WSNs. Highly energy-efficient medium access methods could be developed for such networks, using the cluster head as a centralized scheduler, data sink, and relay for the whole cluster. Issues arising in such contexts include ways to achieve intercluster communication, minimizing intercluster interference, and the possibility of having same application-specific clusters in the MAC layer for optimal performance. All these issues need further investigation.

Scalability. Compared to generic WAHNs, WSNs have a larger node count as well as higher density. This should be a critical consideration in designing medium access schemes. Less scalable protocols may cause unbearable overheads when applied to large networks and may cause extreme energy drains in certain nodes, even causing network failure. On the other hand, quality of service degradations can be severe. In this context, the scalability of currently available MAC schemes should be further investigated.

Mobility management. In certain scenarios, several mobile nodes may be roaming the region of deployment of a WSN among stationary nodes already organized under a certain hierarchy for communication. Sometimes the mobile nodes might serve as gateways, sinks, or localization devices, requiring their proper attachment to certain stationary points of the network. This problem is addressed partially in the SMACS/EAR algorithm described in Section 7.5.2; however, the EAR algorithm does not ensure the optimal use of resources. Thus, further investigation is required in this regard.

Hardware constraints. It is argued that energy savings can often be improved using FDMA or CDMA scheduling schemes, for example, as in SMACS [17]. However, the complexity of the required radio interface poses challenges due to compact hardware of sensor nodes. Use of such schemes must be done in conjunction with a suitable TDMA scheme for energy savings because nodes must listen to the channel all the time under pure FDMA or CDMA schemes. Nevertheless, with possible future improvements in hardware fabrication, such hybrid schemes might have a potential to play a greater role in energy savings while giving superior throughput and delay performance. The main challenge in such schemes is to optimize use of resources, for example, time slots and frequency bands. Transmission ranges of nodes are relatively lower in WSNs, so frequency reuse may be possible in hierarchical, cluster-based WSNs as in traditional cellular networks. Moreover, frequency reuse might require nodes to listen only to a limited number of channels, making such hardware more feasible.

Comparison metrics. While novel medium access schemes for optimum energy savings are developed, due attention should be paid to the metrics used in comparing these schemes. Often a trade-off takes place between energy savings and network performance. Thus, unified metrics should be used during comparisons or this will lead to unfair conclusions and probable confusion. The total energy savings of a WSN depend on percentage sleep time and average length of sleep interval. If used alone, percentage sleep time does not account for the possible higher frequency of switching that may drain a significant amount of node energy. Average sleep length is a preferred metric because it can account for the node switching. Appropriate benchmarks should be developed to facilitate accurate comparison of metrics among different MAC schemes.

7.7   Conclusions

Unique features of WSNs in comparison with generic WAHNs were identified in this chapter. Four prominent ad hoc network medium access methods were briefly discussed, as well as their merits, drawbacks, and implications toward WSNs. Design challenges in MAC for WSNs were emphasized with a classification of application and communication types in WSNs. Four medium access schemes recently proposed toward energy savings in WSNs were discussed, comparing their merits and drawbacks. In addition to these four schemes, a few other medium access schemes that have been recently proposed for WSNs [23–26] were not discussed here to preserve brevity of this chapter.

Finally, several open issues related to energy aware MAC protocol design for WSNs were emphasized. Several research efforts are already underway in this area and are being tested on open source platforms like MICA/TinyOS [27]. Significant research efforts are still required to address these open issues in order to achieve the ultimate objective of energy optimal medium access in sensor networks. Also, it is anticipated that such research efforts will soon lead to open standards for WSNs similar to the currently available, commercially deployed standards for WAHNs.

References

1. I.F. Akyildiz, W. Su, Y. Sankarasubramaniam, and E. Cayirci, Wireless sensor networks: a survey, *Computer Networks*, 38(4), 393–422, 2002.
2. The Institute of Electrical and Electronics Engineers, Wireless LAN medium access control (MAC) and physical layer (PHY) specifications, IEEE Standard 802.11, June 1997.
3. J.C. Haartsen, The Bluetooth radio system, *IEEE Personal Commun. Mag.*, 28–36, Feb. 2000.
4. Bluetooth SIG Inc., Specification of the Bluetooth system: core, http://www.bluetooth.org, 2001.
5. C.E. Jones, K.M. Sivalingam, P. Agrawal, and J.C. Chen, A survey of energy efficient network protocols for wireless networks, *Wireless Networks*, 7(4), 343–358, 2001.

6. K.M. Sivalingam, J.-C. Chen, P. Agrawal, and M. Srivastava, Design and analysis of low-power access protocols for wireless and mobile ATM networks, *Wireless Networks*, 6(1), 73–87, 2000.

7. S. Singh and C.S. Raghavendra, PAMAS: power aware multi-access protocol with signaling for ad hoc networks, *ACM Computer Commun. Rev.*, 28(3), 5–26, July 1998.

8. P. Karn, MACA — a new channel access method for packet radio networks, in *Proc. ARRL/CRRL Amateur Radio 9th Computer Networking Conf.*, 1, 134–140, 1990.

9. V. Bhargawan et al. MACAW: a media access protocol for wireless LANs, *Proc. ACM Sigcomm '94*, 24(4), 212–225, 1994.

10. L. Kleinrock and F. Tobagi, Packet switching in radio channels: carrier sense multiple access modes and their throughput delay characteristics, *IEEE Trans. Commun.*, COM-23(12), 1400–1416, Dec. 1975.

11. Y. Tseng, C. Hsu, and T. Hsieh, Power-saving protocols for IEEE 802.11-based multi-hop ad hoc networks, in *Proc. IEEE Infocom*, 1, 200–209, New York, June 2002.

12. G.J. Pottie and W.J. Kaiser, Wireless integrated network sensors, *Commun. ACM*, 43(5), 51–58, May 2000.

13. Crossbow Inc., Expected battery life vs. system current usage and duty cycle URL: www.xbow.com/Support/Support_pdf_files/PowerManagement.xls, Energy specifications for MICA motes, 2003.

14. Crossbow Inc., Data sheet for MICA2 wireless measurement system, 2003.

15. W. Ye, J. Heidemann, and D. Estrin, An energy-efficient MAC protocol for wireless sensor networks, in *Proc. IEEE Infocomm*, 3, 1567–1576, New York, June 2002.

16. S.R. Madden, M.J. Franklin, J.M. Hellerstein, and W. Hong, The design of an acquisitional query processor for sensor networks, in *Proc. (SIGMOD'03)*, 1, 491–502, San Diego, CA, June 2003.

17. K. Sohrabi, J. Gao, V. Ailawadhi and G. Pottie, Protocols for self-organization of a wireless sensor network, *IEEE Personal Commun. Mag.*, 7(5), 16–27, Oct. 2000.

18. V. Rajendran, K. Obraczka, and J.J. Garcia–Luna–Aceves, Energy-efficient, collision-free medium access control for wireless sensor networks, in *Proc. ACM SIGMOBILE Int. Conf. Embedded Networked Sensor Systems (SenSys 2003)*, 1, 181–192, Los Angeles, CA, November 2003.

19. S. Coleri, PEDAMACS: power efficient and delay aware medium access protocol for sensor networks, M.S. Thesis, Department of Electrical Engineering and Computer Science, University of California, Berkeley, December 2002.

20. L. Bao and J.J. Garcia–Luna–Aceves, A new approach to channel access scheduling for ad hoc networks, in *Proc. IEEE MOBICOMM 2001*, 1, 210–221, Rome, 2001.

21. R. Min, M. Bhardwaj, S.-H. Cho, N. Ickes, E. Shih, A. Sinha, A. Wang, and A. Chandrakasan, Energy-centric enabling technologies for wireless sensor networks, *IEEE Wireless Communications*, 9(4), 28–39, August 2002.

22. E. Shih, S.-H. Cho, N. Ickes, R. Min, A. Sinha, A. Wang, and A. Chandrakasan, Physical layer driven protocol and algorithm design for energy-efficient wireless sensor networks, in *Proc. MOBICOMM 2001*, 1, 272–287, Rome, 2001.

23. A. Woo and D. Culler. Transmission control scheme for media access in sensor networks, in *Proc. MOBICOMM 2001*, 1, 221–235, Rome, 2001.

24. R. Kannan, R. Kalidindi, S.S. Iyengar, and V. Kumar. Energy and rate based MAC protocol for wireless sensor networks, *ACM SIGMOD Record*, Special section on sensor network technology and sensor data management, 32(4), 60–65, December 2003.

25. K. Arisha, M. Youssef, and M. Younis, Energy-aware TDMA-based MAC for sensor networks, in *Proc. IEEE Integrated Manage. Power Aware Commun., Computing Networking (IMPACCT 2002)*, New York City, May 2002.

26. J.M. Van-Dam, An adaptive energy-efficient MAC protocol for wireless sensor networks, M.S. thesis, Delft University of Technology, June 2003.

27. University of California, Berkeley, TinyOS homepage. URL http://webs.cs.berkeley.edu/tos/index.html, 2003.

8

Techniques to Reduce Communication and Computation Energy in Wireless Sensor Networks

Vishnu Swaminathan
Duke University

Yi Zou
Duke University

Krishnendu Chakrabarty
Duke University

8.1 Introduction

Energy consumption is an important design consideration for wireless sensor networks. These networks are useful for a number of applications, such as environment monitoring, surveillance, and target detection and localization. The sensor nodes in such applications operate under limited battery power; they also tend to be situated at remote and/or inaccessible locations and thus the cost of replacing battery packs is high.

Figure 8.1 illustrates the basic structure of a sensor node. In sensor nodes, design for low energy entails energy-efficient sensing, computation (local information processing), and communication. Many strategies have been proposed that target the energy consumption of these functions in sensor networks. Node-level energy minimization techniques target different components, such as the CPU and I/O devices, in each sensor node. Commonly used node-level energy minimization techniques are based on

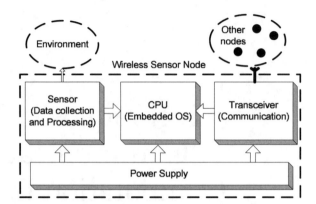

FIGURE 8.1 Typical architecture of a sensor node.

dynamic voltage scaling (DVS) for reducing processor energy and I/O-based dynamic power management (DPM) techniques for reducing the energy consumption of I/O devices.

On the other hand, techniques for energy-efficient communication address energy reduction by significantly reducing unnecessary wireless network traffic. The transmission of detailed sensed information consumes a significant amount of energy due to the large volume of sensed data. In sensor networks for target detection, localization, and classification, energy can be reduced by intelligently querying a select set of nodes for detailed target information without compromising other objectives such as coverage, reliability, etc. This reduces the traffic on the network, thus reducing the energy expended in transmitting large volumes of redundant data.

8.2 Overview of Node-Level Energy Management

One approach to reduce energy consumption is to employ low-power hardware design techniques [7, 14, 33]. These design approaches are static in that they can only be used during system design and synthesis. Thus, these optimization techniques do not fully exploit the potential for node-level power reduction under changing workload conditions and their ability to trade off performance with power reduction is thus inherently limited. An alternative and more effective approach to reducing energy in embedded systems and sensor networks is based on *dynamic power management* (DPM), in which the operating system (OS) is responsible for managing the power consumption of the system.

Many wireless sensor networks are also designed for *real-time* use. Real-time performance is defined in terms of the ability of the system to provide real-time temporal guarantees to application tasks that request such guarantees. These systems must therefore be designed to meet functional as well as timing requirements [6]. Energy minimization adds a new dimension to these design criteria. Thus, although energy minimization for sensor networks is of great importance, energy reduction must be carefully balanced against the need for real-time responsiveness.

Recent studies have shown that the CPU and the I/O subsystems are major consumers of power in an embedded system; in some cases, hard disks and network transceivers consume as much as 20% of total system power in portable devices [18, 25]. Consequently, CPU-centric and I/O-centric DPM techniques have emerged at the forefront of DPM research for wireless sensor networks.

8.2.1 CPU-Centric DPM

Designers of embedded processors used in sensor nodes now include variable-voltage power supplies in their processor designs, i.e., the supply voltages of these processors can be adjusted dynamically to trade off performance with power consumption. *Dynamic voltage scaling* (DVS) refers to the method by which quadratic savings in energy is obtained through the run-time variation of the supply voltage to the processor.

It is well known that the power consumption of a CMOS circuit exhibits a cubic dependence on the supply voltage V_{dd}. However, the execution time of an application task is proportional to the sum of the gate delays on the critical path in a CMOS processor. Because gate delay is inversely proportional to V_{dd}, the execution time of a task increases with decreasing supply voltage. The energy consumption of the CMOS circuit, which is the product of the power and the delay, therefore exhibits a quadratic dependence on V_{dd}.

In embedded sensor nodes, where peak processor performance is not always necessary, a drop in the operating speed (due to a reduction in operating voltage) can be tolerated in order to obtain quadratic reductions in energy consumption. This forms the basis for DVS; the quadratic dependence of energy on V_{dd} has made it one of the most commonly used power reduction techniques in sensor nodes and other embedded systems. When processor workload is low, the OS can reduce the supply voltage to the processor (with a tolerable drop in performance) and utilize the quadratic dependence of power on voltage to reduce energy consumption.

8.2.2   I/O-Centric DPM

Many peripheral devices possess multiple power states — usually one high-power working state and at least one low-power sleep state. Hardware-based timeout schemes for power reduction in such I/O devices have been incorporated into several device designs. These techniques shut down devices when they have been idle for a period of time specified previously. A device that has been placed in the sleep state is powered up when a new request is generated.

With the introduction of the ACPI standard in 1997, the OS was provided with the ability to switch device power states dynamically during run time, thus leading to the development of several new types of DPM techniques. Predictive schemes use various system parameters to estimate the lengths of idle periods for devices. Stochastic models with different probabilistic distributions have been used to estimate the times at which devices can be switched between power states. The goal of these methods, however, is to minimize the response times of devices. Indeed, many such probabilistic schemes see widespread use in portable and interactive systems such as laptop computers. However, their applicability in sensor systems, many of which require real-time guarantees, is limited due to a drawback inherent to probabilistic methods.

Switching between device power states incurs a time penalty, i.e., a device takes a certain amount of time to transition between its power states. In hard real-time systems in which tasks have firm deadlines, device switching must be performed with caution to avoid the potentially disastrous consequences of missed deadlines. The uncertainty inherent in probabilistic estimation methods precludes their use as effective device-switching algorithms in hard real-time systems whose behavior must be predictable with a high degree of confidence. Current-day practice consists of keeping devices in real-time systems powered up during the entirety of system operation; the critical nature of I/O devices operating in real-time prohibits shutting down devices during run time.

This chapter describes two node-level energy reduction methods for wireless sensor networks. The first algorithm focuses on reducing processor energy and was implemented on a laptop equipped with an AMD Athlon 4 processor and running the real-time Linux (RT-Linux) OS. Experiences in implementation are described and experimental power measurements provided; these validate and support simulation results. The second technique targets I/O devices in sensor nodes. The approach includes two online algorithms that schedule the shutdowns and wake-ups for I/O devices in sensor nodes that require hard real-time temporal guarantees.

8.3   Overview of Energy-Efficient Communication

Energy-efficient communication in sensor networks is crucial because most sensor nodes are battery driven and therefore severely energy constrained. Considerable research has been recently carried out in an effort to make communication in sensor networks energy efficient [5, 12, 13, 19, 23, 24, 27, 29, 31, 32, 36, 38, 39]. The focus here is on reducing energy consumption in wireless sensor networks for target localization and data communication. The transmission of detailed target information consumes a significant amount of energy because of the large volume of raw data. Contention for the limited bandwidth among the shared wireless communication channels causes additional delay in relaying detailed target information to the cluster head.

This chapter describes a technique to prolong the lifetimes of the nodes in the sensor network by adopting a new target localization procedure. It details an *a posteriori* energy-aware target localization strategy based on a two-step communication protocol between the cluster head and the sensors reporting the target detection events. In the first step, sensors detecting a target report the event to the cluster head using a very short binary yes/no message. The cluster head subsequently queries a subset of sensors in the vicinity of these likely target positions. This subset is determined through the localization procedure executed by the cluster head. Simulation results show that a large amount of energy is saved by using this procedure. These results also illustrate the built-in advantages of the proposed target localization procedure in reducing communication bandwidth and filtering out false alarms.

8.4   Node-Level Processor-Oriented Energy Management

A set $R = \{r_1, r_2, ..., r_n\}$ of n tasks is given. Associated with each task, $r_i \in R$, are the following parameters: (1) an arrival time a_i; (2) a deadline d_i; and (3) a length l_i (represented as the number of instruction cycles). Each task is placed in the ready queue at time a_i and must complete its execution by its deadline d_i. The tasks cannot be pre-empted. The CPU can operate at one of k voltages: $V_1, V_2, ..., V_k$. Depending on the voltage level, the CPU speed may take on k values: $s_1, s_2, ..., s_k$. The supply voltage to the CPU is controlled by the OS, which can dynamically switch the voltage during run time. The energy E_i consumed by task r_i is proportional to $v_i^2 l_i$. The problem is defined as follows:

P_{cpu}: Given a set R of n tasks and, for each task $r_i \in R$, (1) a release time a_i; (2) a deadline d_i; and (3) a length l_i, and a processor capable of operating at k different voltages, $V_1, V_2, ..., V_k$, with corresponding speeds $S_1, S_2, ..., S_k$, determine a sequence of voltages $v_1, v_2, ..., v_n$ and corresponding speeds $s_1, s_2, ...,$

s_n for the task set R such that the total energy consumed $\left. c_{ij}(x,y) \right\}$ by the task set is minimized, while also attempting to meet as many task deadlines as possible.

8.4.1   The LEDF Algorithm

LEDF is an extension of the well-known earliest deadline first (EDF) algorithm [15], which maintains a list of all released tasks called the ready list. These tasks have an absolute deadline associated with them that is recalculated at each release based on the absolute time of release and the relative deadline. When tasks are released, the task with the earliest deadline is selected for execution. A check is performed to see if the task deadline can be met by executing it at a lower voltage (speed). Each speed at which the processor can run is considered in order from the lowest to the highest. For a given speed, the worst-case execution time of the task is calculated based on the maximum instruction count. If this execution time is too high to meet the current absolute deadline for the task, the next higher speed is considered. Otherwise, a test is applied to verify that all ready tasks will be able to meet their deadlines when the current earliest-deadline task is run at a lower speed.

Procedure LEDF()
t_c: current time;
$S_h > S_{l1} > S_{l2} > \ldots S_{lm}$: Available processor speeds
schedulable = 1
1. **if** ready_list ≠ NULL
2.   Sort task deadlines in ascending order;
3.   Select task τ_i with earliest deadline;
4.   **for** $S = S_{lm}$ to S_h
5.   **if** $t_c + \dfrac{l_i}{S} \le d_i$ **then**
6. $t = t_c + \dfrac{l_i}{S}$
7.   **for** each task τ_u that has not completed execution
8.   **if** $t + \dfrac{l_u}{S_h} \le d_u$ **then**
9. $t = t + \dfrac{l_u}{S_h}$
10.   **else**
11.   schedulable = 0
12.   **break**
13.   **endfor**
14.   **if** schedulable = 0
15.   Schedule t_i at S
16.   **break**
17.   **endif**
18.   **endfor**

FIGURE 8.2 LEDF algorithm.

The test consists of iterating down the ordered list of tasks and comparing the worst-case completion time for each task (at the highest speed) against its absolute deadline. If any task will miss its deadline, the selected speed is insufficient and the next higher speed for the current task is considered. If the deadlines of all tasks in the ready list can be met at the highest speed, LEDF assigns the lower voltage to the task and the task begins execution. When the task completes execution, LEDF again selects the task with the nearest deadline to be executed. As long as tasks are waiting to be executed, LEDF schedules the one with the earliest absolute deadline for execution. Figure 8.2 describes the algorithm in pseudocode form.

For a processor with two speeds, the LEDF algorithm has a computational complexity of $O(n \log n)$ where n is the total number of tasks. The worst-case scenario occurs when all n tasks are released at time $t = 0$. This involves sorting n tasks in the ready list and then selecting the task with the earliest deadline for execution. When more than two speeds are allowed, the complexity of LEDF becomes $O(n \log n + kn)$, where k is the number of speed settings allowed.

8.4.2 Implementation Testbed

8.4.2.1 Hardware Platform

The power measurement experiments were conducted on a laptop with an AMD Mobile Athlon 4 processor. AMD's PowerNow! technology offers greater flexibility in setting frequencies and core voltages [2]. The 1.1-GHz Mobile Athlon 4 processor can be set at several core voltage levels ranging from 1.2 to 1.4 V in 0.05-V increments. For each core voltage, there is a predetermined maximum clock frequency. The power states chosen to use in the scheduler and simulations are shown in

TABLE 8.1   Speed and Voltage Settings for Athlon 4 Processor

Power State	Speed (MHz)	Voltage (V)
1	1100	1.4
2	900	1.35
3	700	1.25

Source: From V. Swaminathan, et al., *Proc. IEEE Real-Time and Embedded Technology and Applications Symposium*, 229–239, 2002. With permission.

Table 8.1. Although only three speeds are used in these experiments, an extension to using all five available speeds appears to be quite straightforward.

PowerNow! technology was developed primarily to extend battery life on mobile systems. Therefore the experiments were conducted on a laptop system rather than a desktop PC. Instead of inserting a current probe into the laptop, system power was simply measured during the experiments. The laptop's system power is drawn from the power converter at approximately 18.5 V DC. Instead of using an oscilloscope or digital ammeter to take exact CPU power measurements at very high frequencies, a simpler approach used a large capacitor to average out the DC current drawn by the entire laptop. This method works primarily because of the periodic nature of these tests.

In a periodic real-time system, the power drawn over one hyperperiod is roughly the same as the power drawn over the next hyperperiod as long as no tasks are added or removed from the task set. Because a fairly large amount of energy needs to be sourced and sunk by the capacitor at the different processor speeds and activity levels, a 30-V DC 360-mF capacitance (160- and 200-mF capacitors in parallel) was used. This capacitance proved capable of averaging current loads for power state periods ranging up to hundreds of milliseconds. (When the processor power state switches at a lower rate than this, the current measurements taken between the AC/DC converter and the voltmeter readings fluctuate.) Figure 8.3 illustrates the experimental hardware setup.

8.4.2.2   Software Architecture

RT-Linux [28] was used as the OS for the experiments. In addition to providing real-time guarantees for tasks and a periodic scheduling system, RT-Linux provides a well-documented method of changing the scheduling policies. An elegant modular interface allows for easy adaptation of the scheduler module to use

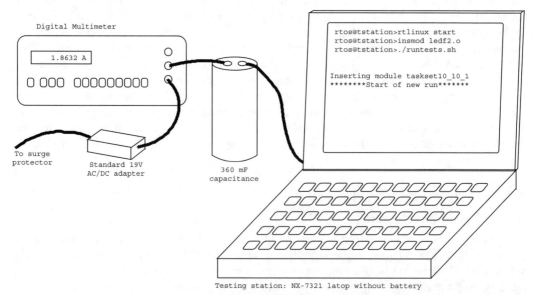

FIGURE 8.3   The experimental setup. (From V. Swaminathan, et al., *Proc. IEEE Real-Time and Embedded Technology and Applications Symposium*, 229–239, 2002. With permission.)

LEDF and then load and unload it as necessary. This feature of RT-Linux was used to swap LEDF for a regular EDF scheduler during power comparisons. Furthermore, RT-Linux uses Linux as its idle task, providing a very convenient method of control and evaluation for the execution of the real-time tasks.

LEDF sorts all tasks by their absolute deadlines and chooses the task with the earliest deadline first. If no real-time tasks are pending, the Linux/Idle task is chosen and run at the lowest available speed. A timeout is then set to preempt the Idle task at the next known release time. Once a speed is identified for a task, the switching code is invoked if the processor is not already operating at that speed.

Switching the power state of a mobile Athlon 4 processor simply consists of writing to a model-specific register (MSR). The core voltage and clock frequency at which the processor is to be set are encoded into a 32-bit word along with 3 control bits. Another 32-bit word contains the stop-grant timeout count (SGTC), which represents the number of 100-MHz system clocks during which the processor is stalled for the voltage and frequency changes. The maximum phase-locked loop (PLL) synchronization time is 50 μs and the maximum time for ramping the core voltage appears to be 100 μs. Calling the WRMSR macro then instruments the power state change. For debugging, the RDMSR macro was used with a status MSR to retrieve the processor's power state. Decoding the two 32-bit word values reveals the maximum, current, and default frequency and core voltage.

The RT-Linux high-resolution timer used for scheduling is based (in x86 systems) on the time-stamp counter (TSC), a special counter introduced by Intel that simply counts clock periods in the CPU since it was started (boot-time). The *gethrtime*() RT-Linux method (and all methods derived from it) convert the TSC value into a time value using the recorded clock frequency. Thus, a simple calculation to determine time in nanoseconds from the TSC value would be the product of TSC and clock period.

RT-Linux was initially developed without the need for dynamic frequency switching, so the speed used for the calculation of time is set at boot time and never changed. Thus, when the processor is slowed to a low-power state with a lower clock frequency, the TSC counts at a lower rate. However, the *gethrtime* method is oblivious to this and the measurement of time slows down proportionally. It is not clear what happens to the TSC, and thus how to measure time, during a speed switch. The TSC does appear to be incremented during some part of the speed switch, but the count is not a reliable means of measuring time. Recalibrating the rate at which the TSC is incremented appears to be a nontrivial task that requires extensive rewriting of the RT-Linux timing code. Therefore, it was decided to track time from within the LEDF module.

8.4.3 Experimental Results

Data from the power measurement experiments, in which total system power consumption of the laptop was measured, are now presented. Knowledge of CPU power savings, however, is useful in generalizing the results. CPU power savings can easily be derived from a set of experiments. In order to isolate the power used by the processor and system board, one can turn off all system components except the CPU and system board, and then take a power reading when the CPU is halted. This power measurement represents the total system power excluding CPU power. This base power is then subtracted from all future power readings in order to obtain CPU power alone. However, halting a processor is far more complex than simply issuing an "HLT" instruction. Decoupling the clock from the CPU involves handshaking between the CPU and the Northbridge. Documentation was not sufficient to implement this.

As an alternative method of estimating power drawn by the system board and components, the power consumption of the CPU with maximum load can be calculated from system measurements at two power states. This can be done by devising tests to isolate power drawn by the LCD screen, hard drive, and the portion of the system beyond control. Once an estimate for system power is available, it can be eliminated from all readings to get an approximation of the fraction of CPU power saved.

Ratios for power consumption in different states can be calculated using the well-known relationship for CMOS power consumption, i.e., $P = f\, a\, C\, V_{dd}^2$, where P is the power; f is the frequency of operation; a is the average switching activity; C is the switching capacitance; and V_{dd} is the operating voltage. The switching capacitance and average switching activity are constant for the same processor and software, so only the frequency and the square of the core voltage are considered. It is also reasonable to assume that other components of the laptop (the screen and hard disk, for example) draw approximately the same current regardless of the CPU operating voltage. Therefore, power state 2 uses approximately 76% as much power as power state 1 and power state 3 uses only 50.7% as much power as the maximum power state. The minimum power configuration for this processor is 300 MHz at 1.2 V, which consumes only 20% of the power consumed in the maximum power state.

In this case, the decision was to compare a fully loaded processor operating at 700 MHz (with a core frequency of 1.25 V) and at 1100 MHz (with a core voltage of 1.4 V). The 700-MHz configuration uses $(700 * 1.25^2)/(1100 * 1.4^2)$, or 50.73% as much CPU power as the 1100-MHz configuration. For a given task running at 1100 MHz, the observed current consumption was 2.373 A. For the same task running at 700 MHz, a current reading of 1.647 A was observed. Assuming that the current consumption of the other components was approximately the same during both runs, the difference in CPU current consumption is 0.726 A. This means that:

$$I_{1100} - I_{700} = 0.726 \Rightarrow I_{1100} - 0.5073 I_{1100} = 0.726 \Rightarrow I_{1100} = 1.474 A$$

In other words, a measured difference of $(2.373 - 1.647) = 0.726$ A of current implies that the fully loaded CPU operating at 1100 MHz draws approximately 1.474 A. Knowing this, it can be deduced from the information in Table 8.2 that the system board and basic components draw approximately 0.456 A, and that under normal operation, the system (including the disk drive and display) draws about 0.976 A in addition to the load from the CPU. This estimation, although approximate, provides a useful method of isolating energy used by the CPU for various utilizations and scheduling algorithms.

Several experiments were performed with three different versions of the scheduling algorithm and different task sets at various CPU utilization levels. A pseudorandom task generator was constructed to generate the test sets. Using the task generator, several random sets of tasks were created. The release times of the tasks are set to the beginning of a period and deadlines to the end. Computation requirements for the tasks are chosen randomly and then scaled to meet the target utilization.

The test programs consist of multiple threads that execute "for" loops for specified periods of time. The time for which these threads run can be determined by examining the assembly level code for each iteration of a loop. Each loop consists of five assembly language instructions, which take one cycle each to execute. The random task set generator takes this into account when generating the task sets.

The simulator is a simple PERL program that reads in task data and generates the schedule that would be generated by the LEDF scheduler. It then takes user-supplied baseline power measurements

TABLE 8.2   Current Consumptions of Various System Components

CPU (1100 MHz)	Screen	Disk	Current Drawn (A)
Idle	Off	STBY	1.5
Idle	Off	On	1.54
Idle	On	STBY	1.91
Idle	On	Sleep	1.9
Idle	On	On	1.97
Max load	Off	STBY	1.93
Max load	On	On	2.45

Source: From V. Swaminathan, et al., *Proc. IEEE Real-Time and Embedded Technology and Applications Symposium*, 229–239, 2002. With permission.

and uses them to compute the power consumption of the task set. Summing up the fraction of the period spent in each state and multiplying it by the appropriate power consumption measurement produces the overall power consumption for the task set. As a reasonable representation of the load generated by the Linux/Idle task, the simulator assumes this task to consume a certain amount of power whose value lies between the power consumptions of a fully loaded and fully idle system running at a given speed. This power value was determined by measuring the power consumption of the laptop with regular Linux running a subset of daemon processes in the background.

A single power-state version of LEDF (in effect, EDF) was used as a comparison point. These tests show the maximum power requirements for the amount of work (computation) to be done. Two- and three-speed versions of LEDF were used to observe the effect of adding additional power states. The two-speed version used operating frequencies of 700 and 1100 MHz, and the three-speed version incorporated an intermediate 900 MHz operating frequency. The CPU utilizations ranged from 10 to 80% in increments of 10%. The maximum utilization of 80% was necessary to guarantee that the Linux/Idle task had sufficient time available for control operations. Without forcing the scheduler to leave 20% of the period open for the Linux/Idle task, the shell became unresponsive, forcing a hard reboot of the machine between each test. The cycle-conserving EDF (ccEDF) algorithm from Pillai and Shin [26] was also implemented and the algorithm compared to it. This implementation of ccEDF uses a set of discrete speeds.

The results are shown in Figure 8.4 for a 15-task task set. Each data point represents the average of three randomly generated task sets for a given utilization value and task set size. LEDF2 (LEDF3) and ccEDF2 (ccEDF3) refer to the use of two (three) processor speeds.

The power savings ranged from 9.4 W in a minimally utilized system to 2.6 W in a fully utilized system. The fully utilized system has lower power consumption under LEDF because LEDF schedules the non-real-time component at the lowest speed. Note, however, that up to the 50% mark the power

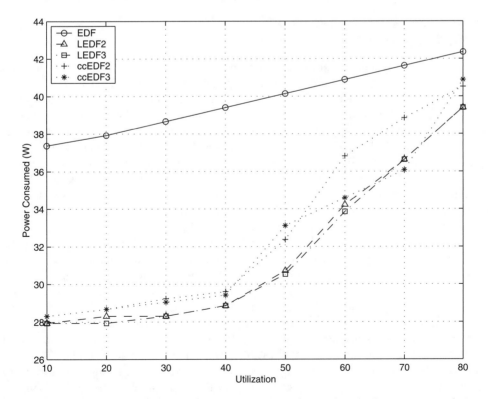

FIGURE 8.4 Heuristic comparison for 15-task task set. (From V. Swaminathan, et al., *Proc. IEEE Real-Time and Embedded Technology and Applications Symposium*, 229–239, 2002. With permission.)

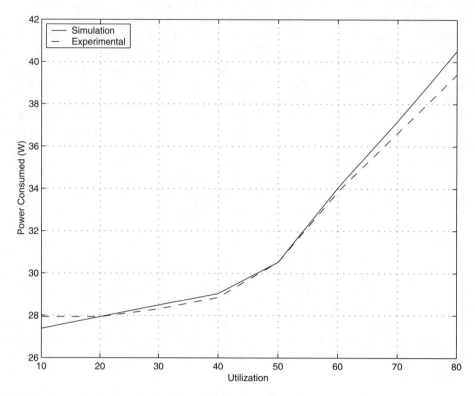

FIGURE 8.5 Comparison of experimental three-state LEDF with expected results. (From V. Swaminathan, et al., *Proc. IEEE Real-Time and Embedded Technology and Applications Symposium*, 229–239, 2002. With permission.)

savings remain over 9 W and, in most cases, remain over 7 W for 60% utilization. With a maximum utilization of 80%, the system can still save significant power with a reasonable task load.

A comparison between measured experimental results and simulation results is shown in Figure 8.5. In most cases, the simulated and measured values are the same or within 2% of each other. The simulation results thus provided a very close match to the experimental results, indicating that the simulation engine model accurately models the real hardware. Because the simulation engine does not take into account the scheduler's computation time, the fidelity of the results may degrade for very high task counts due to the extra cost of sorting the deadlines. In order to verify this, it was decided to evaluate LEDF with several randomly generated task sets with different utilizations, with the number of tasks ranging from 10 to 200, and measure the execution time of the scheduler for each task set.

Results show that the execution time of the scheduler was in the order of microseconds, while the task execution times were in the order of milliseconds. For increasing task set size, scheduler runtime increases at a very slow rate. Thus, scheduling overhead does not prove to be too costly for the power-aware version of EDF. For task sets with more than 240 tasks, the RT-Linux platform tended to become unresponsive. These results are shown in Table 8.3. The entries in the table correspond to task sets with 40% utilization, but with varying numbers of tasks. The other task sets experimented with (task set utilizations of 50 and 80%) also exhibit the same trend in scheduler runtime and are not reproduced here. The scheduler overhead in Table 8.3 indicates the time taken by the scheduler to sort the task set and to identify the active task. Even though implementation of LEDF is currently of $O(n^2)$ complexity and can be replaced by a faster $O(n \log n)$ implementation, it is obvious from the table that scheduling overhead is negligible for over a hundred tasks for utilizations ranging from 10 to 80%. For small task sets in which the task set consists of a few hundred tasks, scheduling overhead is negligible compared to task execution times.

TABLE 8.3 Measured Scheduler Overhead for Varying Task Set Sizes

Number of Tasks	Measured Scheduler Overhead (ns)
10	1739
20	1824
30	1924
60	3817
120	6621
180	10916
200	12243

8.5 Node-Level I/O-Device-Oriented Energy Management

Prior work on DPM techniques for I/O devices has focused primarily on scheduling devices in non-real-time systems. The focus of these algorithms is on minimizing user response times rather than meeting real-time task deadlines; therefore, these methods are not viable candidates for use in real-time systems. Because of their inherently probabilistic nature, the applicability of the preceding methods to real-time systems falls short in one important aspect — real-time temporal guarantees cannot be provided. Such methods perform efficiently in interactive systems in which user waiting time is an important design parameter. In real-time systems, minimizing response time of a task does not guarantee that its deadline will be met. Thus, new algorithms that operate in a deterministic manner are needed in order to ensure real-time behavior.

8.5.1 Device Scheduling for Two-State I/O Devices

This subsection describes LEDES, a deterministic device-scheduling algorithm for two-state I/O devices. It begins by defining the device scheduling problem P_{io}, and describing the assumptions in greater detail.

Take a task set $T = \{\tau_1, \tau_2, \ldots, \tau_n\}$ of n tasks. Each task $\tau_i \in T$ is defined by (1) an arrival time a_i; (2) a worst-case execution time c_i; (3) a period p_i; (4) a deadline d_i; and (5) a device-usage list L_i. The device-usage list L_i for a task τ_i is defined as the set of I/O devices used by τ_i. The hyperperiod H of the task set is defined as the least common multiple of the periods of all tasks. Without loss of generality, assume that the deadline of a task is equal to its period, i.e., $p_i = d_i$.

A set $K = \{k_1, k_2, \ldots, k_p\}$ of p I/O devices is used in the system. Each device k_i is characterized by:

- Two power states: a low-power sleep state $ps_{l,i}$ and a high-power working state $ps_{h,i}$
- A wake-up time from $ps_{l,i}$ to $ps_{h,i}$ represented by $t_{wu,i}$
- A shutdown time from $ps_{h,i}$ to $ps_{l,i}$ represented by $t_{sd,i}$
- Power consumed during wake-up $P_{wu,i}$,
- Power consumed during shutdown $P_{sd,i}$
- Power consumed in the working state $P_{w,i}$
- Power consumed in the sleep state $P_{s,i}$

Requests can be processed by the devices in the working state only. All I/O devices used by a task must be powered up before the task starts execution. Because an *online* device scheduler must be fast and efficient, assume that device scheduling decisions are made only at task starts and completions. Although software timers can potentially be used to switch device power states at *any* time instant, the processing of each timer interrupt incurs an architecture-dependent service-time penalty. This penalty, and therefore its inclusion in the task model and device scheduling algorithms, requires special handling; forcing all device switching to be performed only at task starts and completions is a simpler approach with no significant impact on energy savings.

Device stays powered up for
entire interval t_{be}

Device performs two transitions
during interval t_{be}

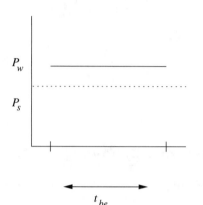

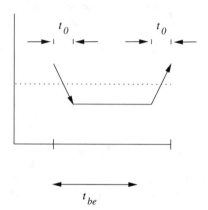

FIGURE 8.6 The time interval for which the energy consumptions are the same in (a) and (b) is called breakeven time. (From V. Swaminathan and K. Chakrabarty, *IEEE Transactions on Computer-Aided Design of Integrated Circuits & Systems*, 22, 847–858, July 2003. With permission.)

In I/O devices, the power consumed by a device in the sleep state is less than the power consumed in the working state, i.e., $P_{s,i} < P_{w,i}$. Without loss of generality, assume that for a given device, k_i, $t_{wu,i} = t_{sd,i} = t_{0,i}$ and $P_{wu,i} = P_{sd,i} = P_{0,i}$. The energy consumed by device k_i is given by:

$$E_i = P_{w,i}t_{w,i} + P_{s,i}t_{s,i} + MP_{0,i}t_{0,i},$$

where M is the total number of state transitions for k_i; $t_{w,i}$ is the total time spent by device k_i in the working state; and $t_{s,i}$ is the total time spent in the sleep state.

Incorrectly switching power states can cause increased, rather than decreased, energy consumption for an I/O device. Incorrect switching of I/O devices is eliminated using the concept of *breakeven time* [14], which is defined as the time interval for which a device in the powered-up state consumes an energy exactly equal to the energy consumed in shutting a device down, leaving it in the sleep state and then waking it up (Figure 8.6). If any idle time interval for a device is greater than the breakeven time t_{be}, energy is saved by shutting it down. For idle time periods that are less than the breakeven time, energy is saved by keeping it in the powered-up state.

Also assume that the start time for each job is known *a priori*. Several commercial RTOS support tick-driven scheduling and mission-critical embedded systems require an inherently deterministic scheduling mechanism [20]. In such types of systems, task schedules are generated offline, and the start times of all jobs are known prior to run time.

The device scheduling problem τ_2 is defined as:

- P_{io}: Given the start times $S = \{s_1, s_2, \ldots, s_n\}$ of the n tasks in a real-time task set T that uses a set K of I/O devices, determine a sequence of sleep/working states for each I/O device $v_s \in G$ such that the total energy consumed, $\Sigma_{i=1}^{p} E_i$, by K is minimized and all tasks meet their respective deadlines.

The following sections describe the conditions under which device state transitions are allowed to minimize energy and ensure the timely completion of tasks. These conditions are different for different scenarios; the scenarios are dependent on the execution times of the tasks that comprise the task set and the number of sleep states present in a device. Begin by assuming that all task execution times are greater than the maximum transition time among all devices and all devices have only one sleep state. It will then be shown that when devices have multiple power states, ensuring timeliness becomes more complex.

One notable advantage of online I/O device scheduling is that online DPM decision making can exploit underlying hardware features such as buffered reads and writes. A device schedule constructed offline and stored as a table in memory precludes the use of such features due to its inherently deterministic approach. The flexibility of online scheduling enhances the effectiveness of device scheduling.

The need for deterministic I/O device scheduling policies is motivated in detail in Swaminathan and Chakrabarty [34], who showed that it is not possible to ensure timely completion of tasks without *a priori* knowledge of future device requests. A naive, probabilistic algorithm cannot be used for real-time task sets. The determinism required to make device-scheduling decisions in hard real-time systems can be quantified through the notion of *look-ahead*, which is a bound on the number of tasks whose device-usage lists must be examined before making a state transition decision, in order to guarantee that no task deadline is missed. Next, the low-energy device scheduler (LEDES) is presented for online scheduling of I/O devices with two power states.

8.5.1.1   Online Scheduling of Two-State Devices: Algorithm LEDES

LEDES assumes that the execution times of all tasks are greater than the transition times of the devices they use. Under this assumption, the amount of look-ahead required before making wake-up decisions to ensure timeliness is easily bounded. This result is derived by presenting the following theorem from Swaminathan and Chakrabarty [34]:

Theorem 8.1. Given a task schedule for a set T *of n tasks with completion times* c_1, c_2, ..., c_n; *the device utilization for each task; and an I/O device* k_j, *it is necessary and sufficient to look ahead m tasks to guarantee timeliness, where m is the smallest integer such that* $\sum_{i=1}^{m} c_i \geq t_{0j}$.

In most practical cases, the completion times of tasks are greater than the transition times $t_{0,i}$ of device k_j. This leads to the following corollary to Theorem 8.1:

Corollary 8.1. Given a task schedule for a set T *of tasks with completion times* c_1, c_2, ..., c_n; *the device utilization for each task; and an I/O device* k_j, *it is necessary and sufficient to look ahead one task to ensure timeliness if the completion times of all tasks in* T *are greater than the transition time* t_{0j} *of device* k_j.

The LEDES algorithm operates as follows (also see Figure 8.7). At the start of task τ_i (line 1), devices not used by the next "immediate" tasks τ_i and τ_{i+1} are put in the sleep state (lines 3 and 4). The time difference between the start of τ_{i+1} and the end of τ_i's execution is evaluated and compared with the transition time $t_{0,j}$ to determine whether k_j's wake-up can be guaranteed at τ_i's finish time. If k_j is powered down, then a wake-up decision must be made (line 8). A device must be waked up at s_i if its wake-up cannot be deferred to τ_i's finish time. This is implemented in line 12 and the device is waked up if needed.

If the scheduling instant at which LEDES is invoked is the completion time of τ_i (line 11) and if k_j is powered up (line 12), it can be shut down only if it can enter the powered-down state fully before s_{i+1} because it may be necessary for it to wake up again. If k_j is in the sleep state (line 15) and is used by τ_{i+1}, it must be waked up to ensure the timely start of τ_{i+1}. These decisions are made for each device and the entire process repeats at each scheduling instant. (Although no mention is made of the break-even time in Figure 8.7, an implicit check is made to ensure that the idle period for a given device is always greater than the breakeven time.)

A simple extension to LEDES can efficiently schedule devices that possess multiple sleep states with the ability to switch from any low-power state directly to the working state. Such a device can be viewed as a device with only two power states. Although the transition times from the sleep states to the powered-up state (and vice-versa) may be different, the correct sleep state to switch a device to is

Algorithm LEDES (k_j, τ_i, τ_{i+1})

1.   **if** curr = s_i
2.   **if** k_j is powered-up
3.   **if** $k_j \notin L_i \cup L_{i+1}$
4.   **shutdown** k_j
5.   **if** $k_j \in L_{i+1}$
6.   **if** $s_{i+1} - (s_i + c_i) \geq t_0, j$
7.   **shutdown** k_j
8.   **else**
9.   **if** $k_j \in L_{i+1}$ and $s_{i+1} - (s_i + c_i) < t_0, j$
10.   **wakeup** k_j
11. **if** curr = $s_i + c_i$
12.   **if** k_j is powered-up
13.   **if** $k_j \notin L_{i+1}$ and $s_{i+1} - $ curr $\geq t_0, j$
14.   **shutdown** k_j
15.   **else**
16.   **wakeup** k_j

FIGURE 8.7   LEDES algorithm.

identified simply by performing a series of transition-time checks to verify that there is sufficient time to wake the device up if it is switched to the selected sleep state. However, LEDES cannot make full use of the available sleep states for devices which possess multiple sleep states, but do *not* possess the ability to jump to any sleep state from the powered-up state.

We next present a more general I/O-centric power management algorithm for hard real-time systems. This algorithm is called the multistate constrained low energy scheduler (MUSCLES). MUSCLES can also schedule devices without the ability to jump from the powered-up state to any sleep state. Therefore, it can be assumed that at a device scheduling instant, a device may be switched from one power state to the next higher or lower power state, i.e., only a single transition is possible at any scheduling instant. The next section describes the MUSCLES algorithm in greater detail.

8.5.2   Low-Energy Device Scheduling of Multistate I/O Devices

The properties of a real-time periodic task remain unchanged from Section 8.5.1. However, I/O device properties now include parameters to describe the different power states. These device properties are restated here for the sake of completeness. Each I/O device $k_i \in K$ is now characterized by:

- A set $PS_i = \{ps_{i,1}, ps_{i,2}, ..., ps_{i,m}\}$ of m sleep states
- A powered-up state $ps_{i,u}$
- Transition time from $ps_{i,j}$ to $ps_{i,j-1}$, denoted by $t_{wu}^{i,j}$
- Transition time from $ps_{i,j}$ to $ps_{i,j+1}$, denoted by $t_{sd}^{i,j}$
- Power consumed during switching up from state $ps_{i,j}$ to $ps_{i,j-1}$, denoted by $P_{wu}^{i,j}$
- Power consumed during switching down from state $ps_{i,j}$ to $ps_{i,j+1}$, denoted by $P_{sd}^{i,j}$
- Power consumed in the working state P_w^i
- Power consumed in sleep state $ps_{i,j}$, denoted by $P_s^{i,j}$,

Assume, without loss of generality, that for each device $k_j \in K$; $t_{wu+1}^{i,j} = t_{sd}^{i,j} = t_{0,i}$; and $P_{wu}^{i,j+1} = P_{sd}^{i,j} = P_{0,i}$. The total energy E_i consumed by device k_i over the entire hyperperiod is given by

$$E_i = P_w^i t_w^i + \sum_{j=1}^{m} P_s^{i,j} t_s^{i,j} + M P_{0,i} t_{0,i},$$

where M is the number of state transitions; t_w^i is the total time spent by the device in the working state; and $t_s^{i,j}$ is the total time spent by the device in sleep state $ps_{i,j}$. In order to provide conditions under which devices can be shut down and powered up, a few important terms are first defined.

Intertask time. The intertask time IT_i for task τ_i is the time interval between the start of task τ_{i+1} and completion of task τ_i. Thus, $IT_i = s_{i+1} - (s_i + c_i)$. Two scheduling instants are associated with a task τ_i. These correspond to the start and completion time of τ_i, respectively. For minimum-energy device scheduling under real-time constraints, it is not always possible to schedule devices at all scheduling instants. This is formalized using the notion of a valid scheduling instant.

Valid scheduling instant. The completion time of τ_i is defined to be a valid scheduling instant for device k_j if $s_{i+1} - (s_i + c_i) \geq t_{0,j}$. In other words, the completion time of τ_i is a valid scheduling instant if and only if $IT_i \geq t_{0,j}$. The start time of τ_i is always a valid scheduling instant. Thus, a task, τ_i, can have one or two scheduling instants, depending on the magnitude of IT_i relative to the transition time $t_{0,j}$ of a device k_j. Valid scheduling instants are important for energy minimization. Wake-ups can be scheduled at these points to minimize energy and also ensure that real-time requirements are met. Consider the example shown in Figure 8.8. This figure shows two tasks, τ_i and τ_{i+1}, with the intertask time $IT_i < t_{0,j}$. Assume that device k_1 (first used by task τ_{i+2}) is in state $ps_{1,1}$ at τ_i's completion time $(s_i + c_i)$. If a device were to be waked up at $s_i + c_i$, it would complete its transition to state $ps_{1,0}$ only in the middle of τ_{i+1}'s execution and would be in the higher powered state for the rest of τ_{i+1}'s execution (i.e., until the next scheduling instant). If, on the other hand, the device were to be waked up at s_{i+1}, one can still ensure that the device is powered-up before task τ_{i+2} starts (with the assumption that $c_{i+1} > t_{0,1}$). However, the device stays in the lower powered state until s_i, resulting in greater energy savings. Thus, wake-ups at valid scheduling instants always result in lowered energy consumption. It is always preferable to wake a device up as late as possible in order to utilize the full potential of online device scheduling.

Subsection 8.5.1 showed that a look-ahead of one task is sufficient when devices have only one sleep state. However, a look-ahead of one task is not sufficient when devices have multiple low-power sleep states. This is clarified through the example shown in Figure 8.9, which shows the execution of three tasks, τ_1, and τ_2, τ_3. Assume that the start time of τ_1 is the current scheduling instant and assume that tasks τ_1 and τ_2 do not use device k_2, which is in sleep state $ps_{2,2}$ at time s_1. An algorithm using a look-ahead of one task, i.e., looking ahead only to task τ_2, would erroneously decide that it is not necessary to wake up k_2 at time s_1. The same situation arises at scheduling instant $s_i + c_i$. At τ_2's start time (s_2), looking ahead to task τ_3, k_2 is switched to state $ps_{2,1}$. At τ_2's completion time, again looking ahead one task to τ_3, k_2 is switched up to the powered-up state $ps_{2,u}$. However, if the intertask time IT_2 were less than $t_{0,2}$, k_2 would not have sufficient time to wake up, resulting in τ_3, missing its deadline.

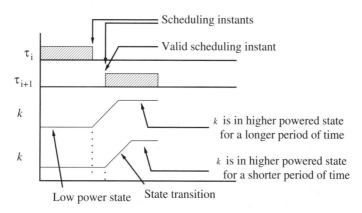

FIGURE 8.8 An invalid scheduling instant. (From V. Swaminathan and K. Chakrabarty, *IEEE Transactions on Computer-Aided Design of Integrated Circuits & Systems*, 22, 847–858, July 2003. With permission.)

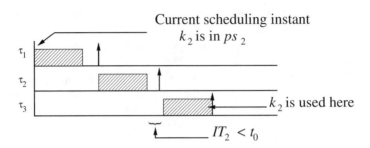

FIGURE 8.9 Look-ahead of one task is insufficient when devices have multiple sleep states. (From V. Swaminathan and K. Chakrabarty, *IEEE Transactions on Computer-Aided Design of Integrated Circuits & Systems*, 22, 847–858, July 2003. With permission.)

From this example, it is interesting to note that look-ahead represented as the number of future *tasks* is inadequate for devices with multiple low-power states. When devices have multiple states, look-ahead must be represented as the number of *valid scheduling instants* between tasks. In fact, the notion of look-ahead changes slightly when considering multiple-state I/O devices. Scheduling complexity thus increases with increasing look-ahead due to the additional computational burden of determining look-ahead; thus, minimizing look-ahead makes the scheduler more efficient. An upper bound on the look-ahead necessary to ensure timeliness while making shut-down decisions for a device [34] is now presented.

Theorem 8.2 Consider an ordered set $T = \{\tau_1, \tau_2, ..., \tau_n\}$ *of n tasks that have been scheduled a priori. Let* c_1, $c_2, ..., c_n$ *be the set of p I/O devices used by the tasks in T. In order to decide whether to switch a device* k_i $\in$ K *from state* $ps_{i,j}$ *to* $ps_{i,j+1}$ *at task* τ_{i+1}*'s start or completion time, it is necessary and sufficient to look ahead* L *tasks, where* L *is the smallest integer such that the total number of valid scheduling instants associated with the sequence of tasks* $\tau_c, \tau_{c+1}, ..., \tau_{c+L-1}$, *excluding the current scheduling instant, is at least equal to j + 1. The device* k_i *can be switched down from* $ps_{i,j}$ *to* $ps_{i,j+1}$ *if no task* τ_t, $c \leq t \leq c + L - 1$, *uses device* k_i.

If the intertask times of all tasks are less than the transition time $t_{0,j}$ for device k_j, Theorem 8.1 yields the following corollary.

Corollary 8.2 Suppose the intertask time IT_i *is less than the transition time* $t_{0,i}$ *for every task* $\tau_c \in T$. *In order for a device* $k_j \in K$ *to be switched down from state* $ps_{i,j}$ *to* $ps_{i,j+1}$ *at the start or completion time of task* τ_c, *it is necessary and sufficient to look ahead j + 1 tasks to ensure timeliness. Moreover, no task* τ_t, $i \leq t \leq j$, *must use device* k_i.

On the other hand, if the intertask times for all tasks are greater than or equal to the transition time $t_{0,j}$, Theorem 8.2 leads to the following corollary.

Corollary 8.3. Suppose the intertask time IT_i *is greater than or equal to the transition time* $t_{0,j}$ *for every task* τ_{i+1}. *In order for a device* $k_j \in K$ *to be switched down from state* $ps_{i,j}$ *to* $ps_{i,j+1}$ *at the start or completion time of task* τ_c, *it is necessary and sufficient to look ahead* $\frac{j+1}{2}$ *tasks to ensure timeliness. Moreover, device* k_j *must not be used by any task* τ_t, $i \leq t \leq j$.

Look-ahead increases as the depth of the sleep state increases. Next, an upper bound on look-ahead for making wake-up decisions is presented.

Theorem 8.3. Consider an ordered set $T = \{\tau_1, \tau_2, ..., \tau_n\}$ *of n tasks and a set* $K = \{k_1, k_2, ..., k_p\}$ *of p devices used by the tasks in T. Suppose the first task after* τ_c *that uses device* k_i *is* τ_{c+L}. *The device* $k_i \in K$ *must be switched up from state* $ps_{i,j+1}$ *to* $ps_{i,j}$ *at the start or completion time of task* τ_c *if and only if the total number of valid scheduling instants, including the current scheduling instant, associated with the tasks* $\tau_c, \tau_{c+1}, ..., \tau_{c+L-1}$ *is exactly equal to j + 1, where* L *is the look-ahead from the current scheduling instant.*

Algorithm MUSCLES (*S*, *PS*, k_i)
curr: current scheduling instant;
At s_m:
1. Find first task τ_L that uses device k_i;
2. Compute number of valid scheduling instants X between s_m and τ_L;
3. **if** $X \geq j + 1$
4.   **switchdown** k_i from $ps_{i,j}$ to $ps_{i,j+1}$;
5. **else if** $X = j$
6.   **wake up** k_i from $ps_{i,j}$ to $ps_{i,j-1}$;
At $s_m + c_m$:
7. Find first task τ_L that uses device k_i;
8. Compute number of valid scheduling instants X between s_m and τ_L;
9. **if** $X \geq j + 1$
10.   **switchdown** k_i from $ps_{i,j}$ to $ps_{i,j+1}$;
11. **else if** $X = j$ and curr is a valid scheduling instant
12.   **wake up** k_i from $ps_{i,j}$ to $ps_{i,j-1}$;
13. **else** leave k_i in $ps_{i,j}$.

FIGURE 8.10   MUSCLES algorithm.

Theorem 8.2 and Theorem 8.3 form the basis for the MUSCLES algorithm described in the next subsection.

8.5.2.1   Online Scheduling for Multistate Devices: Algorithm MUSCLES

For a precomputed task schedule, MUSCLES generates a sequence of power states for every device so that energy is minimized. It operates as follows (also see Figure 8.10): Let device *ki* be in state $ps_{i,j}$ at scheduling instant s_m. MUSCLES finds the next task, τ_L, that uses k_i (line 1). A check is then performed to test whether k_i can be switched down to a lower powered state. This is done by ensuring that at least $j + 1$ valid scheduling instants are between the current scheduling instant and τ_L's start time. The presence of $j + 1$ valid scheduling instants implies that device k_i can be switched down from state $ps_{i,j}$ to $ps_{i,j+1}$ (line 3). The absence of $j + 1$ valid scheduling instants precludes the shutting down of k_i to a lower powered state; a check is then performed to test whether the device must be switched up. If exactly j instants are present, then the device must be switched up in order to ensure timeliness (line 4). At the completion of a task, τ_m, the same process is repeated. However, in order to minimize energy consumption, an additional check is performed to test if the current scheduling instant is valid. If the current scheduling instant is not valid, the device is left in the same state until a valid scheduling instant (line 10) occurs. MUSCLES guarantees that no task ever misses its deadline.

LEDES and MUSCLES are polynomial-time algorithms. MUSCLES has a worst-case complexity of $O(pn^2)$, where *p* is the number of I/O devices used in the system and *n* is the number of tasks in the task set; LEDES is $O(p)$. The complexity increases in MUSCLES because the amount of look-ahead, in terms of valid scheduling instants, for each device must be computed before any state transition. Nevertheless, the relatively low complexity of MUSCLES makes online device scheduling for low-energy and real-time execution feasible.

8.5.3   Experimental Results

LEDES and MUSCLES were first evaluated with several randomly generated task sets with varying utilizations. The task sets consist of six tasks with varying hyperperiods and randomly generated device-usage lists. Because jobs may be preempted, each preempted slice of a job is considered as two jobs with identical device-usage lists. As a result, the number of jobs listed in each task set in Table 8.4 is an approximation. Each task in the task set uses one or more out of three I/O devices whose power values are shown in Table 8.5. These values pertain to real devices currently deployed

TABLE 8.4   Evaluation Task Sets for LEDES and MUSCLES

Task Set	Approximate Number of Jobs	Hyperperiod
T_1	303	1,700
T_2	68,951	567,800
T_3	36,591	341,700

Source: From V. Swaminathan and K. Chakrabarty, *IEEE Transactions on Computer-Aided Design of Integrated Circuits & Systems*, 22, 847–858, July 2003. With permission.

TABLE 8.5   Device Parameters Used in Evaluation of LEDES and MUSCLES

Device k_i	Device Type	P_w	$P_{sd}^{i,0} = P_{wu}^{i,1}$	$P_{sd}^{i,1} = P_{wu}^{i,2}$	$P_{sd}^{i,2} = P_{wu}^{i,1}$	t_0	$P_s^{i,1}$	$P_s^{i,2}$	$P_s^{i,3}$
k_1	HDD[a]	2.3 W	1.5 W	0.6 W	0.3 W	0.6 s	1.0 W	0.5 W	0.2 W
k_2	NIC[b]	0.3 W	0.2 W	0.05 W	—	0.5 s	0.1 W	3 mW	—
k_3	DSP[c]	0.63 W	0.4 W	0.1 W	—	0.5 s	0.25 W	0.05 W	—

[a] Fujitsu MHL2300AT Hard Disk Drive. http:://www.fujitsu.jp/hypertext/hdd/drive/overseas/mhl2xxx/mhl2xxx.html.
[b] AMD Am79C874 NetPHY-1LP Low-Power 10/100 Tx/Rx Ethernet Transceiver Technical Datasheet.
[c] Analog Devices Multiport Internet Gateway Processor.*http://www.analog.com.*
Source: From V. Swaminathan and K. Chakrabarty, *IEEE Transactions on Computer-Aided Design of Integrated Circuits & Systems*, 22, 847–858, July 2003. With permission.

in embedded systems. Each task set is scheduled using the rate-monotonic algorithm. The utilization of each task set is varied from 10 to 90% to observe the impact of slack on energy consumption of the I/O devices.

While evaluating LEDES, it was assumed that the single low-power sleep state for the devices corresponded to the highest powered sleep state of the device. The energy consumptions at different utilizations for task set T_1 are shown in Figure 8.11. Figure 8.12 illustrates the percentage energy savings for each of the task sets obtained from the LEDES algorithm.

A study of Figure 8.11 reveals that the energy consumption using LEDES and MUSCLES increases with increasing utilization because devices are kept powered up for longer periods of time within the hyperperiod. The resulting decrease in sleep time causes this increased energy consumption. However, energy savings of over 35 and 40% can be obtained for task sets with high and low utilization, respectively. No task deadlines are missed at any utilization value.

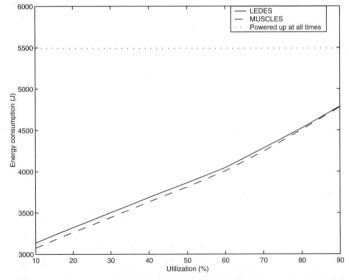

FIGURE 8.11   Comparison of LEDES and MUSCLES for task set T_1. (From V. Swaminathan and K. Chakrabarty, *IEEE Transactions on Computer-Aided Design of Integrated Circuits & Systems*, 22, 847–858, July 2003. With permission.)

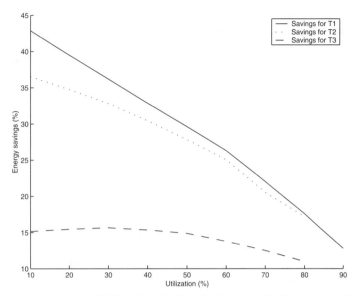

FIGURE 8.12 Energy savings using LEDES. (From V. Swaminathan and K. Chakrabarty, *IEEE Transactions on Computer-Aided Design of Integrated Circuits & Systems*, 22, 847–858, July 2003. With permission.)

One other important observation is that the savings in energy obtained from MUSCLES *over* LEDES decreases with increasing utilization because the number of valid scheduling instants decreases with increasing utilization. Thus, MUSCLES cannot place devices in deep sleep states as often in high-utilization task sets as it can in low-utilization task sets.

LEDES and MUSCLES were also evaluated with three real-life task sets. These task sets are used in an instrument navigation system (INS) [16]; a computer numerical control (CNC) system [17]; and an aviation platform (GAP) [21]. The assignment of devices to tasks in the task sets has been inferred from the functionality of the tasks. For example, task 2 in the GAP task set is a communication task that uses the NIC and task 7 is a status update task that performs occasional reads and writes and therefore uses a hard disk.

Table 8.6 presents the energy consumptions for these task sets using LEDES and MUSCLES. The energy values here are expressed in units of joules and correspond to the energy consumption of the I/O devices over the duration of a single hyperperiod. Using LEDES, an energy savings of 45% for the GAP task set are obtained. With MUSCLES, an energy savings of 80% is obtained for the INS task set. Owing to the low utilizations of real-life task sets, significant energy savings can be obtained by intelligently performing state transitions for I/O devices.

8.6   Energy-Aware Communication

This section describes a novel target localization approach based on a two-step communication protocol between the cluster head and the sensors within the cluster. Because the energy consumption in wireless sensor networks increases significantly during periods of activity, which may be triggered, for example, by a moving target [5], an energy-reduction method is proposed for target localization

TABLE 8.6   Comparison of LEDES and MUSCLES Using Real-Life Task Sets

Task Set	Energy (J)			% Savings	
	All Powered Up	LEDES	MUSCLES	(LEDES)	(MUSCLES)
CNC	403,104	197,140	117,604	51%	70%
INS	16.5×106	7.7×106	3×106	51%	81%
GAP	381×106	210×106	153×106	45%	60%

Source: From V. Swaminathan and K. Chakrabarty, *IEEE Transactions on Computer-Aided Design of Integrated Circuits & Systems*, 22, 847–858, July 2003. With permission.

in cluster-based wireless sensor networks. In the first step, sensors detecting a target report the event to the cluster head. The amount of information transmitted to the cluster head is limited. In order to conserve power and bandwidth, the message from the sensor to the cluster head is kept very small; in fact, the presence or absence of a target can be encoded in just one bit. No detailed information such as signal strength, confidence level in the detection, imagery or time series data is transmitted at this time. Based on the information received from the sensor and knowledge of the sensor deployment within the cluster, the cluster head executes a probabilistic scoring-based localization algorithm to determine the likely position of the target. It subsequently queries a subset of sensors in the vicinity of the likely target position.

8.6.1 Detection Probability Table

The cluster head first generates a detection probability table for each grid point. Consider a sensor field represented by an $m \times n$ grid. Let $\langle (y, x), (y, x + 1) \rangle$ denote the set of deployed sensor nodes and $|S| = k$. Let s be an individual sensor node on the sensor field located at grid point (x,y). Sensor detections are imprecise, so coverage is expressed in probabilistic terms. For any grid point P at (i,j), the coverage $c_{ij}(x,y)$ of P by a sensor located at $sp = (x,y)$ is expressed probabilistically in Equation 8.1, which is motivated in part by Elfes [9]:

$$c_{ij}(x,y) \begin{cases} 0, & \text{if } r + r_e \leq d_{ij}(x,y) \\ e^{-\lambda a^\beta}, & \text{if } r - r_e < d_{ij}(x,y) < r + r_e \\ 1, & \text{if } r - r_e \geq d_{ij}(x,y) \end{cases} \tag{8.1}$$

where $r_e(r_e < r)$ is a measure of the uncertainty in sensor detection; $a = d_{ij}(x,y) - (r - r_e)$; and λ and β are parameters that measure detection probability when a target is at distance greater than r_e but within a distance from the sensor. This model reflects the behavior of range sensing devices such as infrared and ultrasound sensors. The probabilistic sensor detection model is shown in Figure 8.13. Note that distances are measured in units of grid points. This figure also illustrates the translation

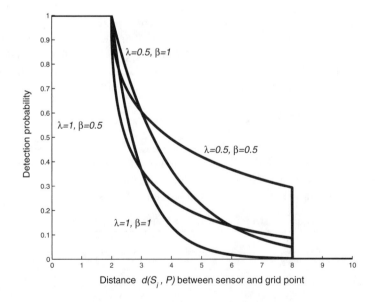

FIGURE 8.13 Probabilistic sensor detection model. (From Y. Zou and K. Chakrabarty, *ACM Transactions on Embedded Computing Systems*, 3, 61–91, February 2004. With permission.)

TABLE 8.7   Example Probability Table

l	$d_1 d_2 d_3$	$p\_table_{ij}(l),\ 0 \le l < 2^{k_{ij}},\ k_{ij} = 3$
0	000	$(1 - 0.5736) \times (1 - 1) \times (1 - 0.5736) = 0.0$
1	001	$(1 - 0.5736) \times (1 - 1) \times 0.5736 = 0.0$
2	010	$(1 - 0.5736) \times 1 \times (1 - 0.5736) = 0.1819$
3	011	$(1 - 0.5736) \times 1 \times 0.5736 = 0.2446$
4	100	$0.5736 \times (1 - 1) \times (1 - 0.5736) = 0.0$
5	101	$(1 - 0.5736) \times (1 - 1) \times 0.5736 = 0.0$
6	110	$0.5736 \times 1 \times (1 - 0.5736) = 0.2446$
7	111	$0.5736 \times 1 \times 0.5736 = 0.3290$

Source: From Y. Zou and K. Chakrabarty, *ACM Transactions on Embedded Computing Systems*, 3, 61–91, February 2004. With permission.

of a distance response from a sensor to the confidence level as a probability value about this sensor response. Different values of the parameters α and β yield different translations reflected by different detection probabilities, which can be viewed as the characteristics of various types of physical sensors.

The detection probability table contains entries for all possible detection reports from sensors that can detect a target at this grid point. Assuming that the sensor field is represented by an $m \times n$ grid, and a grid point P at (i, j) is covered by a set of k_{ij} sensors denoted as S_{ij}, $|S_{ij}| = k_{ij}, 0 \le k_{ij} \le k$ and $S_{ij} \subseteq \{s_1, s_2, \ldots, s_k\}$.

The probability table is built on the power set of S_{ij} because there are 2^k_{ij} possibilities for k_{ij} sensors in reporting an event. These 2^k_{ij} cases include the case that none of the sensors detect anything (represented by the binary string as "00...0") as well as the case that all of the sensors (represented by the binary string as "11...1") detect an event. Thus the probability table for grid point (i, j) then contains 2^k_{ij} entries, defined as:

$$p\_table_{ij}(l) = \prod_{s_p \in S_{ij}} p_{ij}(S_p, l) \tag{8.2}$$

where $0 \le l \le 2^k_{ij}$, and $p_{ij}(sp, l) = c_{ij}(sp)$ if s_j detects a target at grid point $P(i, j)$; otherwise, $p_{ij}(sp, l) = 1 - c_{ij}(sp)$. Table 8.7 gives an example of the probability tables on a 5×5 grid with three sensors deployed.

Consider a grid point P that is covered by three sensors, s_1, s_2 and s_3, with probabilities as 0.57, 1, and 0.57, respectively.[*] For these three sensors, eight possibilities exist for their combined event detection at P. For example, the binary string 110 denotes the possibility that s_1 and s_2 report a target but s_3 does not report a target. For each such possibility $d_1 d_2 d_3$ ($d_1, d_2, d_3 \in \{0, 1\}$ for a grid point, the conditional probabilities that the cluster head receives $d_1 d_2 d_3$, given that a target is present at that grid point, are calculated. Table 8.7 lists these conditional probabilities for this example. Consider the binary string 110, the conditional probability associated with this possibility, is given by $p\_table_{24}(6) = p_{24}(s_1, 6)\, p_{24}(s_2, 6)\, p_{24}(s_3, 6) = 0.57 \times 1 \times (1 - 0.57) = 0.24$. Note that the probability table generation is only a one-time cost. Once this table is generated, there is no need to refresh it unless sensor locations are changed.

8.6.2   Score-Based Ranking

After the probability table is generated for all the grid points, localization is done by the cluster head if a target is detected by one or more sensors. An inference method based on the established probability table is used. When time instant t the cluster head receives a positive event message from $k(t)$ sensors, it uses the grid point probability table to determine which of these sensors are most suitable to be queried for more detailed information. Detailed target reporting consumes more

---

[*]These coverage values can be obtained using the sensor detection model described in Zou and Chakrabarty [37].

energy consumption and needs more bandwidth. Therefore, the cluster head cannot afford to query all the sensors for detailed reports. Sensor detection information also has an inherent redundancy, so it is not necessary to query all sensors. The scoring approach is able to select the most suitable sensors for this purpose.

Consider the 10×10 grid shown in Figure 8.14. There are five sensors deployed, $k = 5$, $r = 2$, and $r_e = 1$. The zigzag shaped line is the target movement trace. The target starts to move at $t = t_{start}$ from the grid point marked as "start" and finishes at $t = t_{end}$ at the grid point marked as "end." Figure 8.15 gives the score report at the time instant t_{start} when the target is present at "start."

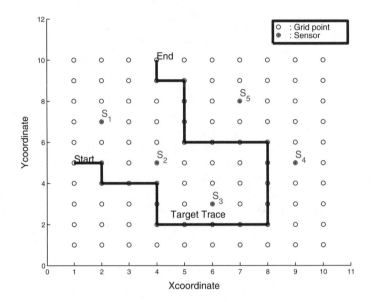

FIGURE 8.14 Sensor field with a moving target. (From Y. Zou and K. Chakrabarty, *ACM Transactions on Embedded Computing Systems*, 3, 61–91, February 2004. With permission.)

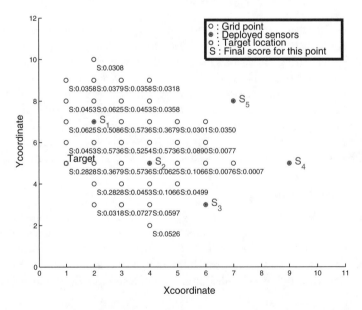

FIGURE 8.15 Scoring results for target in the sensor field at t_{start}. s_1 and s_2 have reported. (From Y. Zou and K. Chakrabarty, *ACM Transactions on Embedded Computing Systems*, 3, 61–91, February 2004. With permission.)

Assume $S_{rep}(t)$ is the set of sensors that have reported the detection of an object at time t and $S_{rep,ij}(t)$ is the set of sensors that can detect a target at point $P(i,j)$ and have also reported the detection of an object at time t. Obviously, $S_{rep,ij}(t) \subseteq S_{rep}(t)$ and $S_{rep,ij}(t) \subseteq S_{ij}$ because $S_{rep,ij}(t) = S_{rep,ij}(t) \cap S_{ij}$. The score of the grid point $P(i,j)$ at time instant t is calculated as follows:

$$SCORE_{ij}(t) = p\_table_{ij}(l) \times w_{ij}(t) \tag{8.3}$$

where l is the index of the p_table_{ij}. The parameter l is calculated from S_{ij} and $S_{rep,\ ij}$. The parameter $p\_table_{ij}\ (l(t))$ corresponds to the conditional probability that the cluster head receives this event information if there was a target at $P(i, j)$. The weight $w_{ij}(t)$ reflects the confidence level in this reporting event for this particular grid point. In previous work [37], the authors have used the weight factor

$$w_{ij}(t) = \frac{k_{rep,ij}(t)}{k_{rep}(t)}$$

where $k_{rep}(t) = |S_{rep}(t)|$, and $k_{rep,\ ij}(t) = |S_{rep,ij}(t)|$; this is sufficient for selecting sensors in order to conserve energy. However, in order to refine the grid point scores to narrow down grid points that are most probably close to the current target location, $w_{ij}(t)$ have been redefined here to improve the accuracy for target location. The weight for the grid point $P(i, j)$ at time instant t is defined as

$$w_{ij}(t) = \begin{cases} 0 & \text{if } S_{rep,ij}(t) = \{\varnothing\} \\ 4^{-\Delta k_{rep,ij}(t)} & \text{otherwise} \end{cases} \tag{8.4}$$

where $\Delta k_{rep,ij}(t)$ measures the degree of difference in the set of sensors that reported, and those that can detect, point $P(i,j)$ at time instant t. The parameter $\Delta k_{rep,ij}\ (t)$ is defined as

$$\Delta k_{rep,ij}(t) = |k_{rep}(t) - k_{rep,ij}(t)| + |k_{rep}(t) - k_{ij}| \tag{8.5}$$

The parameter $w_{ij}(t)$ is therefore a decaying factor that is 1 only if $S_{rep}(t) = S_{ij}$. The number 4 in the formula for $w_{ij}(t)$ was chosen empirically after it was found to provide accurate simulation results; $w_{ij}(t)$ is used to filter out grid points not likely to be close to the actual target location. The score is based on the probability value from the probability table and the current relationship among $S_{rep}(t)$, $S_{rep,ij}(t)$, and S_{ij}. Table 8.8 gives some score calculation examples for the grid points in Figure 8.15 at the time instant t_{start}.

TABLE 8.8   Scoring Calculation Example for t at t_{start}

(x,y)	S_{ij}	$S_{rep,ij}(t)$	$w_{ij}(t)$	$p_{tableij}\ (l(t))$	$SCORE_{ij}(t)$
...	...	...	...	...	...
(1,6)	s_1	s_1	0.25	0.7248	0.0453
(2,6)	s_1, s_2	s_1, s_2	1.00	0.5736	0.5736
(3,6)	s_1, s_2	s_1, s_2	1.00	0.5254	0.5254
(4,6)	s_1, s_2	s_1, s_2	1.00	0.5736	0.5736
(5,6)	s_2, s_5	s_2	0.25	0.3562	0.0890
(6,6)	s_2, s_3, s_5	s_2	0.25	0.1240	0.0077
...	...	...	...	...	...

8.6.3   Selection of Sensors to Query

Assume that the maximum number of sensors allowed to report an event is k_{max}, and the set of the sensors selected by the cluster head for querying at time t is $S_q(t)$, $S_q(t) \subseteq S_{rep}(t) \subseteq \{s_1, s_2, \ldots, s_k\}$. To select the sensor to query based on the event reports and the localization procedure, first note that, for time instant t, if $k_{max} \geq k_{rep}(t)$, then all reporting sensors can be queried. Otherwise, sensors are selected on a score-based ranking. The sensors selected correspond to the ones that have the shortest distance to grid points with the highest scores. This selection rule is defined as:

$$S_q(t){:}d\Big(S_q(t), P_{MS}\Big) = min\Big\{d\big(s_i, P_{MS}\big)\Big\} \tag{8.6}$$

where $s_i \in S_{rep}(t)$, and P_{MS} denotes the set of grid points with the highest scores.

Note that multiple grid points with the maximum score are possible. When this happens, the score concentration is calculated by averaging the scores of the current grid point and its eight neighboring grid points. The grid point with the highest score (or the score concentration) is the most likely current target location. Therefore, selecting sensors closest to this point guarantees that the selected sensors can provide the most detailed and accurate data in response to the subsequent queries. Target identification is not possible at this stage because the cluster head has no additional information other than $S_{rep}(t)$. However, the selected sensors provide enough information in the subsequent stage to facilitate target identification.

The accuracy of this target localization procedure is evaluated by calculating the distance between the grid point with the highest score and the actual target location. For the example of Figure 8.14, Table 8.9 gives some results for the selected sensor when the target is moving from "start" ($t = 1$) to "end." It is assumed that $k_{max} = 1$ and the target is moving at a constant speed. $\bar{S}_q(t)$ is the set of sensors closest to the actual location of the target at time t. The results show that $S_q(t)$ matches $\bar{S}_q(t)$ in many cases. The example does not illustrate the advantages of this proposed strategy because not many sensors are actually involved at the same time for target detection. However, in Subsection 8.6.6 the proposed algorithm performs very well when many sensors are involved in the target detection and reporting process.

8.6.4   Energy Evaluation Model for Target Localization in Wireless Sensor Networks

Consider the energy consumption for a sensor network that is actively detecting a target in the sensor field. Assume that sensor nodes are homogeneous and therefore the energy consumption for sensing is the same for each sensor node. Because the focus here is on energy minimization of communication traffic due to target activities or events, energy consumed by sensor nodes when they are in the idle

TABLE 8.9   Selected Sensors for the Example in Figure 8.14

t	$S_{rep}(t)$	$\bar{S}_q(t)$	$S_q(t)$	t	$S_{rep}(t)$	$S_q(t)$	$\bar{S}_q(t)$
...	...	...	...	...	...	...	...
3	s_1, s_2	s_1	s_2	4	s_2	s_2	s_2
5	s_2, s_3	s_3	s_2	6	s_2, s_3	s_3	s_2
7	s_2, s_3	s_3	s_3	8	s_3	s_3	s_3
...	...	...	...	...	...	...	...
16	s_4, s_5	s_4	s_4	17	s_4, s_5	s_4	s_5
18	s_2, s_3, s_5	s_2	s_2	19	s_2, s_5	s_5	S_2
20	s_1, s_2, s_5	s_2	s_2	21	s_5	s_5	s_5
...	...	...	...	...	...	...	...

state is not considered. This does not imply, however, that the energy consumption of idle sensor nodes can always be ignored.

To simplify the energy analysis, first consider a primitive sensor model that focuses on the energy consumption of the wireless sensor network due to the target activities or events. Suppose the sensor node has three basic energy consumption types — sensing, transmitting and receiving — and these power values (energy per unit time) are E_s, E_t, and E_r, respectively. If all sensors that reported the target for querying are selected, the total energy consumed for the event happening at time instant t can be evaluated using the following set of equations:

$$E_1(t) = k_{rep}(t)(E_t + E_r)T_1 \tag{8.7}$$

$$E_2(t) = (k_{rep}(t)E_r + E_t)T_2 \tag{8.8}$$

$$E_3(t) = k_{rep}(t)(E_t + E_r)T_3 \tag{8.9}$$

$$E_4(t) = E_s T_s \tag{8.10}$$

$$E(t) = E_1(t) + E_2(t) + E_3(t) + E_4(t) \tag{8.11}$$

$$E = \sum_{t=t_{start}}^{t_{end}} E(t) \tag{8.12}$$

where

E_1 is the energy required for reporting the detection of an object.

E_2 is the energy required for transmitting query information from the cluster head by broadcasting and for receiving this information at the sensor nodes.

E_3 is the energy required by sensor nodes being queried to send detailed information to the cluster head.

Parameters T_1, T_2, and T_3 denote the lengths of time involved in the transmission and reception, which are directly proportional to the sizes of data for yes/no messages, control messages to query sensors, and the detailed sensor data transmitted to the cluster head, respectively.

Parameter T_s is the time of sensing activity of sensors.

Parameter E denotes total energy — in this case for target localization from t_{start} to t_{end}.

For the proposed probabilistic localization approach, the total energy consumption E^* is calculated as follows:

$$E_1^*(t) = k_{rep}(t)(E_t + E_r)T_1 \tag{8.13}$$

$$E_2^*(t) = (k_q(t)E_r + E_t)T_2 \tag{8.14}$$

$$E_3^*(t) = k_q(t)(E_t + E_r)T_3 \tag{8.15}$$

$$E_4^*(t) = E_s T_s \tag{8.16}$$

$$E^*(t) = E_1^*(t) + E_2^*(t) + E_3^*(t) + E_4^*(t) \tag{8.17}$$

$$E^* = \sum_{t=t_{start}}^{t_{end}} E^*(t) \tag{8.18}$$

where $E_1(t)^* = E_1(t)$; $E_4^*(t) = E_4(t)$; and the total energy consumed is denoted by E^*. Therefore, the energy savings via the use of the probabilistic target localization algorithm is:

$$\Delta E = E - E^* = C \sum_{t=t_{start}}^{t_{end}} (k_{rep}(t) - k_q(t)) \tag{8.19}$$

where $C = E_r\, T_2 + (E_t + E_r)T_3$ is a constant. Because $k_q(t)$ is always less than or equal to $k_{rep}(t)$, $\Delta E \geq 0$. Also, ΔE is monotonically nondecreasing with time. Figure 8.16 shows the energy saved for the target trace in Figure 8.14.

8.6.4.1 Refined Energy Evaluation Model

The previous primitive energy evaluation models given by Equation 8.7 through Equation 8.19 ignore the overhead due to the two-step protocol and convey the impression that large volumes of data can greatly burden the energy consumption on sensor nodes. Therefore, the energy evaluation model is refined to incorporate the overhead introduced by this approach. The refined model is used later as the primary metric for evaluating energy consumption with parameter values from Heinzelman et al. [13] and Rappaport [27]. It is still necessary to consider a sensor node with three basic energy consumption types — sensing, transmitting, and receiving — and these power values (joules per second) are ψ_s, ψ_t, and ψ_r, respectively. Assume at time instant t, $k(t)$ sensors have detected the target,

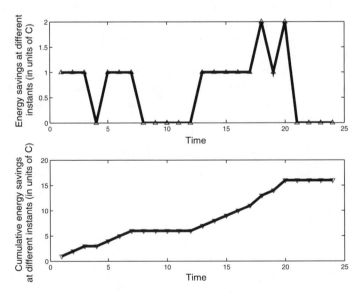

FIGURE 8.16 Energy saved for the example in Figure 8.14 using the primitive energy evaluation model. (From Y. Zou and K. Chakrabarty, *ACM Transactions on Embedded Computing Systems*, 3, 61–91, February 2004. With permission.)

where $k(t) \le k$. Therefore, the energy for sensing activities in the wireless network, denoted as $E_s(k(t))$, is

$$E_s(k(t)) = k(t)\psi_s T_s \tag{8.20}$$

where T_s is the time duration that a sensor node is involved in sensing.

For a fixed time interval, E_s is a constant if all sensor nodes are assumed to be homogenous. The energy used for communication between nodes and the cluster head can be categorized into two types, E_b and E_c. The parameter E_c is the energy consumed by a sensor node for communication with the cluster head. This includes the energy for transmitting data and the energy for receiving data. The parameter E_b is the energy needed for broadcasting data from the head to the nodes. E_b and E_c are functions of T and $k(t)$, where T is the time required for retrieving data from a sensor node or broadcasting data from the cluster head, and $k(t)$ is the number of sensors involved in this communication at time instant t. E_c and E_b are defined as:

$$E_c(k(t), T) = (\psi_t T + \psi_r T)k(t) \tag{8.21}$$

$$E_b(k(t), T) = \psi_t T + \psi_r Tk(t) \tag{8.22}$$

The parameter T is directly proportional to the volume of data involved in the communication. In this work, T can be one of three values: T_d for raw target data; T_e for target event reporting; and T_q for query request. They satisfy the relationship $T_e \le T_q \ll T_d$ because raw data collected by a sensor node can be up to hundreds of bytes in size. Assume that target detection and localization are discrete processes derived from a discrete sampling of target activities in the sensor network. Also, because the sensor network is designed to track target activities, T_s, T_e, T_q, and T_d are assumed to be less than the granularity of the time t. Thus, for the case of a target moving in the sensor field during the time interval $[t_{start}, t_{end}]$, the corresponding instantaneous energy consumption $E(t)$ and total energy consumption E in the wireless sensor network can be expressed as

$$E(t) = E_s(k(t)) + E_c(k(t), T_d) \tag{8.23}$$

$$E = \sum_{t=t_{start}}^{t_{end}} E(t). \tag{8.24}$$

From Equation 8.20 to Equation 8.24, energy consumption is evaluated using the preceding target localization method:

$$E^*(t) = E_s(k_{rep}(t)) + E_c(k_{rep}(t), T_e) + E_b(k_q(t), T_q) + E_c(k_q(t), T_d) \tag{8.25}$$

$$E^* = \sum_{t=t_{start}}^{t_{end}} E^*(t) \tag{8.26}$$

Let $k(t) = k_{rep}(t)$ in Equation 8.2 and Equation 8.3. The difference in energy consumption, $\Delta E = E - E^*$ can be expressed as:

$$\Delta E(t) = (k_{rep}(t) - k_q(t))(\psi_t + \psi_r)T_d$$

$$-(k_q(t)\psi_r + \psi_t)T_q - k_{rep}(t)(\psi_t + \psi_r)T_e \tag{8.27}$$

$$\Delta E = \sum_{t=t_{start}}^{t_{end}} \Delta E(t) \tag{8.28}$$

The last two terms in Equation 8.27 indicate the overhead for the proposed target localization procedure. Because $T_d \gg T_e$ and $T_d \gg T_q$, the overhead is small. Because $k_q < k_{max}$, with k_{max} properly selected, from Equation 8.27 and Equation 8.28, energy consumption is greatly reduced with the passage of time.

8.6.5   Procedural Description

Figure 8.17 shows the pseudocode of the procedure to generate the detection probability table for each grid point and Figure 8.18 shows pseudocode for simulation of the probabilistic localization algorithm. For an $n \times m$ grid with k sensors, the computational complexity involved in generating the probability table is $O(nm2k)$ because the maximum number of sensors that can detect a grid point is k for the worst case. The computational complexity of the localization procedure is $O(nmk_{max})$, $k_{max} \le k$. Therefore, the computational complexity of the probabilistic localization algorithm is $\max\{O(nmk_{max}), O(nm2^k)\} = O(nm2^k)$. Even though the worst-case complexity of the localization procedure is exponential in k, in practice the localization procedure can execute in less time because the number of sensors that effectively detect a target at a given grid point is small.

8.6.6   Simulation Results

This subsection presents results for case studies carried out on a Pentium III 1.0GHz PC using Matlab.

8.6.6.1   Case Study

The simulation is presented on a 30×30 sensor field grid with 20 sensors randomly placed in the sensor field. The parameters of the sensor detection model are $r = 5$; $r_e = 4$; $\lambda = 0.5$; and $\beta = 0.5$.

Procedure *Generate_Probability_Table* (P(i, j), {s_1, ..., s_k})

```
1/*find S_ij, the set of sensors that can detect P(i, j)*/
2 For sp ∈ {s₁, s₂, ..., s_k}
3    if d_ij (s_p) ≤ r + r_e
4       S_ij = S_ij U {s_p}
5    End
6 End
7/*fill up the probability table */
8 For l, 0 ≤ l ≤ k_ij, k_ij = |S_ij|;
9    If s_p detects P(i, j)
10      Set p_ij (s_p, l) = cij (s_p);
11   Else
12      Set p_ij (s_p, l) = 1 – cij (s_p);
13   End
14 Set p_table_ij(l) =   ∏   p_ij (s_p, l)
15 Else              s_p ∈ S_ij
```

FIGURE 8.17 Pseudocode for generating the detection probability table. (From Y. Zou and K. Chakrabarty, *ACM Transactions on Embedded Computing Systems*, 3, 61–91, February 2004. With permission.)

Procedure *Target_Localization*(Grid, {s_1, ..., s_k}, TargetTrace)

/* k_{max} is the maximum number of sensors that are allowed for querying, p_{rep} is the threshold level for a sensor to report to the cluster head of an event. *TargetTrace* starts from t_{start} and ends at t_{end}, with time unit as 1. */

```
1  Set t = t_start;
2  While (t ≤ t_end)
3      /* current target location */
4      Set Target = TargetTrace(t);
5      /* calculate the scores */
6      Calculate S_rep(t) from {s_1, s_2, ..., s_k}, Target(t), p_rep;
7      Set k_rep(t) = |S_rep(t)|;
8      For P(i, j) in Grid, i ∈ [1, width], j ∈ [1, height]
9          Set k_ij = |S_ij|;
10         Calculate S_rep,ij(t) from S_rep(t) and P(i, j);
11         Calculate the index l(t) of p_table_ij from
                S_rep(t), and S_rep,ij(t);
12         Set k_rep,ij(t) = |S_rep,ij(t)|;
13         If S_rep,ij(t) = {ø}
14             w_ij(t) = 0;
15         Else
16             Set Δk_rep,ij(t) = |k_rep(t) − k_rep,ij(t)|
                    + |k_rep(t) − k_ij(t)|;
17             w_ij(t) = 4^(−Δk_rep.ij(t));
18         End
19         Set SCORE_ij(t) = p_table_ij (l(t) × w_ij(t));
20     End
21     /* select sensors for querying */
22     Calculate S_q(t) from SCORE_ij(t) and k_max, i ∈ [1, width], j ∈ [1, height];
23     /* next time instant */
24     Set t = t + 1;
25 End
```

FIGURE 8.18 Pseudocode of the target localization procedure.

Choose the energy consumption model parameters as $\psi_r \approx 400$ nJ/sec; $\psi_t \approx 400$ nJ/sec; and $\psi_s \approx 1000$ nJ/sec. These values are based on typical values given in Heinzelman et al. [13] and Rappaport [27], assuming the sensing rate for the sensor is 8 bits/sec. No physical data are available for T_d and T_e; however, because their values do not affect the target localization procedure, it is necessary only to set them manually to satisfy the relationship $T_d \gg T_e$ and $T_d \gg T_q$. In this case, $T_d = 100$ ms; $T_e = 2$ ms; and $T_q = 4$ ms.

The layout of the sensor field is given in Figure 8.19, with a target trace randomly generated in the sensor field. The target travels from the position marked "start" to the position marked "end." Assume the target locations are updated at discrete time instants in units of seconds, and the granularity of time is long enough for sampling by two neighboring locations in the target trace with negligible errors. Evaluate the algorithm for $k_{max} = 1$; $k_{max} = 2$; and $k_{max} = 3$.

Figure 8.20 illustrates the instantaneous energy savings in percentage, and Figure 8.21 shows the absolute value of the cumulative energy savings for the case study as the target moves along its trace in the sensor field. The energy savings are compiled relative to the base case when all sensors report complete target information in one step everywhere.

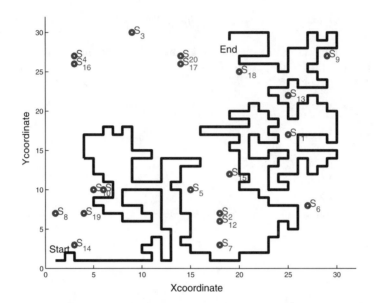

FIGURE 8.19 Sensor field layout with target trace. (From Y. Zou and K. Chakrabarty, *Proc. IEEE International Conference on Pervasive Computing and Communications,* 60–67, 2003. With permission.)

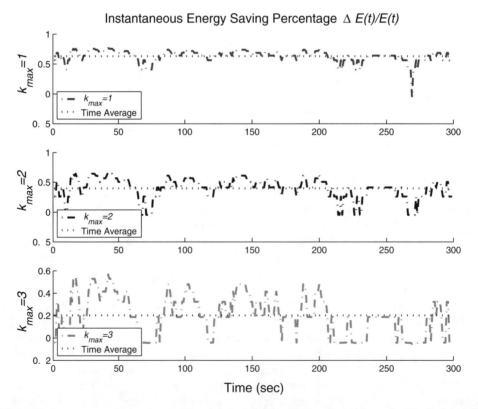

FIGURE 8.20 Instantaneous energy saving percentage during target localization relative to the "always report" one-step base case. (From Y. Zou and K. Chakrabarty, *Proc. IEEE International Conference on Pervasive Computing and Communications,* 60–67, 2003. With permission.)

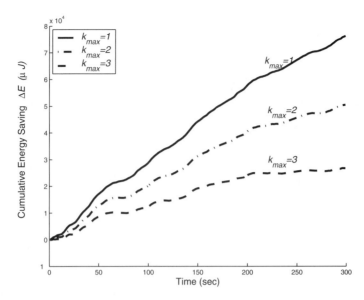

FIGURE 8.21 Cumulative energy saving during target localization relative to the "always report" one-step base case. (From Y. Zou and K. Chakrabarty, *Proc. IEEE International Conference on Pervasive Computing and Communications*, 60–67, 2003. With permission.)

It is evident from Figure 8.20 and Figure 8.21 that a large amount of energy is saved during target localization. Note that when k_{max} approaches $k_{rep}(t)$, the savings is less apparent due to the additional communication overhead of the two-stage query protocol. Nevertheless, a considerable amount of energy is saved in target localization, even when $k_{max} = 3$. With an appropriate selection of k_{max}, the proposed algorithm performs exceptionally well.

Next, consider the latency in the localization of a target by the cluster head. Latency refers to the time taken for the cluster head to collect detailed target information from sensor nodes starting from the time sensor nodes detect an event. Assume that the wireless sensor network uses the time division multiple access (TDMA) protocol [35]. The results are shown in Figure 8.22. The latency is reduced here compared to the base case using a "report once" strategy because a large amount of communication for transmitting raw data has been reduced to a smaller amount of data sent by a selected set of sensors. This is an added advantage to the proposed energy-aware target localization procedure.

Because the selection of sensors for querying is based on the detection probability table and the distance of sensors from the estimated high-score points, the *a posteriori* approach offers another important advantage: it provides a substantial amount of built-in false-alarm filtering. Figure 8.23 illustrates the false-alarm filtering ability of the proposed approach. False alarms reported by some malfunctioning sensors during $t \in [18, 22]$ by s_4; during $t \in [138, 142]$ by s_{16}; and during $t \in [239, 241]$ by s_8 were manually generated. The distance d of the target from the sensor in $S_{rep}(t)$ farthest from it was calculated, as well as the distance d^* of the target from the sensor in $S_q(t)$ farthest from it. The difference $d - d^*$ is used as a measure of the built-in filtering ability. Figure 8.23 shows the variation of $d - d^*$ with time. Note that prior to querying, the cluster head only knows which sensors have reported the detection of a target; no detailed information about the target is available to the cluster head.

The localization approach successfully narrows down the sensors closest to the real target location and selects them for detailed information querying. As shown in Figure 8.23, the three spikes represent the fact that the false alarms from the sensor (which in this case is the sensor farthest from the actual target location) have been filtered out because the proposed target localization procedure is still able to select the most appropriate sensors to be queried for detailed target information.

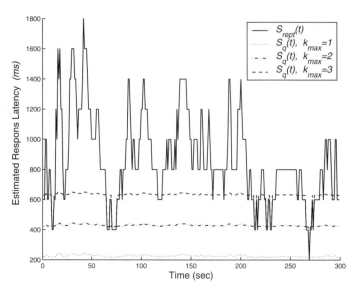

FIGURE 8.22 Latency in the localization of a target by the cluster head. (From Y. Zou and K. Chakrabarty, *Proc. IEEE International Conference on Pervasive Computing and Communications*, 60–67, 2003. With permission.)

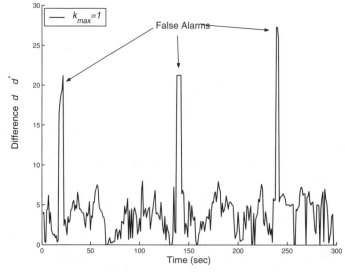

FIGURE 8.23 Results on localization error in the presence of false alarms. (From Y. Zou and K. Chakrabarty, *Proc. IEEE International Conference on Pervasive Computing and Communications*, 60–67, 2003. With permission.)

8.7 Conclusions

Energy is an important resource in battery-operated sensor systems. For such systems that operate under real-time constraints, energy consumption must be carefully balanced with real-time responsiveness. This chapter has described two approaches to energy minimization in sensor networks: node-level energy minimization and network-level energy minimization.

The two node-level energy minimization techniques described here focus on minimizing the energy consumption of the processor and I/O devices in a sensor node, respectively. Implementation of a dynamic power management scheme that uses an EDF-based scheduler to support real-time execution was described. The scheduler is efficient and can be easily integrated into the kernels of real-time operating systems on sensor nodes. The LEDF algorithm provides significant energy savings in real-time systems.

In many embedded systems, the I/O subsystem is a viable candidate to target for energy reduction. Two low-energy I/O device scheduling algorithms have also been described. The first, called LEDES, assumes that the I/O devices present in the sensor system possess two power states: a high-powered working state and a low-powered sleep state. Even under this somewhat restrictive assumption, experimental results show that energy savings of over 40% can be obtained. A generalized version of LEDES, called MUSCLES, that schedules devices with more than two low-power sleep states has also been described. Experimental case studies for real-life task sets show that energy savings of over 50% can be obtained by targeting the I/O subsystem for power reduction. The amount of energy saved decreases with increasing task-set utilization; nevertheless, energy savings of over 40% with these device scheduling algorithms in high-utilization task sets can be realized.

Finally, the chapter described an energy-aware target localization procedure for cluster-based wireless sensor networks for target localization and detection to reduce network-level energy consumption. This approach is based on the combination of a two-step communication protocol between the cluster head and the sensors in the cluster and a probabilistic localization algorithm. This approach reduces energy consumption, decreases the latency for target localization, and provides a mechanism for filtering false alarms.

References

1. AMD Am79C874 NetPHY-1LP Low-Power 10/100 Tx/Rx Ethernet Transceiver Technical Datasheet.
2. AMD PowerNow! Technology, *http://www.amd.com/us-en/Processors/ProductInformation/0,,30_118_756_807964,00.html.*
3. Analog Devices Multiport Internet Gateway Processor.*http://www.analog.com.*
4. L. Benini, A. Bogliolo, G.A. Paleologo, and G. De Micheli. Policy optimization for dynamic power management. *IEEE Trans. Computer-Aided Design,* 16(6), 813–833, June 1999.
5. M. Bhardwaj and A.P. Chandrakasan. Bounding the lifetime of sensor networks via optimal role assignments. *Proc. IEEE Infocom Conf.,* 1587–1596, 2002.
6. G.C. Buttazzo, *Hard Real-time Computing Systems: Predictable Scheduling Algorithms and Applications,* Kluwer Academic Publishers, Norwell, MA, 1997.
7. A.P. Chandrakasan and R. Broderson, *Low Power Digital CMOS Design,* Kluwer Academic Publishers, Norwell, MA, 1995.
8. E.-Y. Chung, L. Benini and G. De Micheli. Dynamic power management using adaptive learning tree. *Proc. Int. Conf. Computer-Aided Design,* 274–279, 1999.
9. A. Elfes. Occupancy grids: a stochastic spatial representation for active robot perception. *Proc. Conf. Uncertainty in AI,* 60–70, 1990.
10. Fujitsu MHL2300AT Hard Disk Drive. http:://www.fujitsu.jp/hypertext/hdd/drive/overseas/mhl2xxx/mhl2xxx.html.
11. R. Golding, P. Bosh, C. Staelin, T. Sullivan and J. Wilkes. Idleness is not sloth. *Proc. Usenix Tech. Conf. UNIX Adv. Computing Syst.,* 201–212, 1995.
12. W.B. Heinzelman, J. Kulik, and H. Balakrishnan, Adaptive protocols for information dissemination in wireless sensor networks, *Proc. IEEE/ACM MobiCom Conf.,* 174–185, 1999.
13. W.B. Heinzelman, A. Chandrakasan and H. Balakrishnan. Energy-efficient communication protocol for wireless micro sensor networks. *Proc. Annu. Hawaii Int. Conf. Syst. Sci.,* 3005–3014, 2000.
14. C. Hwang and A.C.-H. Wu. A predictive system shutdown method for energy saving of event-driven computation. *Proc. Int. Conf. Computer-Aided Design,* 28–32, 1997.
15. K. Jeffay, D.F. Stanat, and C.U. Martel. On non-preemptive scheduling of periodic and sporadic tasks with varying execution priority. *Proc. Real-Time Syst. Symp.,* 129–139, December 1991.
16. D. Katcher, H. Arakawa, and J. Strosnider. Engineering and analysis of fixed priority schedulers. *IEEE Trans. Software Eng.,* 19, 920–934, September 1993.

17. N. Kim, M. Ryu, S. Hong, M. Saksena, C. Choi, and H. Shin. Visual assessment of a real-time system design: case study on a CNC controller. *Proc. Real-Time Syst. Symp.*, 300–310, 1996.
18. K. Li, R. Kumpf, P. Horton and T. Anderson. A quantitative analysis of disk drive power management in portable computers. *Proc. Usenix Winter Conf.*, 279–292, 1994.
19. S. Lindsey and C.S. Raghavendra. PEGASIS: power-efficient gathering in sensor information systems. *Proc. IEEE Aerospace Conf.*, 3, 1125–1130, 2002.
20. J.W.S. Liu. *Real-Time Systems.* Prentice Hall, Upper Saddle River, NJ, 2000.
21. D.C. Locke, D. Vogel and T. Mesler. Building a predictable avionics platform in Ada: a case study. *Proc. Real-Time Syst. Symp.*, 181–189, 1991.
22. Y-H. Lu, L. Benini and G. De Micheli. Operating system directed power reduction. *Proc. Int. Conf. Low-Power Electron. Design*, 37–42, 2000.
23. A. Manjeshwar and D.P. Agrawal. TEEN: a routing protocol for enhanced efficiency in wireless sensor networks. *Proc. Int. Parallel Distributed Process. Symp.*, 2009–2015, 2001.
24. R. Min, M. Bhardwaj, S.H. Cho, A. Sinha, E. Shih, A. Wang, and A. Chandrakasan. Low-power wireless sensor networks. *VLSI Design*, 2001.
25. M. Newman and J. Hong. A look at power consumption and performance of the 3Com Palm Pilot. *http://guir.cs.berkeley.edu/projects/p6/finalpaper.html.*
26. P. Pillai and K.G. Shin. Real-time dynamic voltage scaling for low-power embedded operating systems. *Proc. Symp. Operating Syst. Principles*, 89–102, 2001.
27. T. Rappaport. *Wireless Communications: Principles & Practice*, Englewood Cliffs, NJ: Prentice Hall, Inc., 1996.
28. The RT-Linux Operating System, *http://www.fsmlabs.com/community/.*
29. E. Shih, B.H. Calhoun, H.C. Seong and A.P. Chandrakasan. An energy-efficient link layer for wireless micro sensor networks. *Proc. IEEE Computer Soc. Workshop VLSI*, 16–21, 2001.
30. T. Simunic, L. Benini, P. Glynn, and G. De Micheli. Event driven power management. *IEEE Trans. Computer-Aided Design*, 840–857, 2001.
31. A. Sinha and A. Chandrakasan. Dynamic power management in wireless sensor networks. *IEEE Design Test Computers*, 18, 62–74, 2001.
32. S. Slijepcevic and M. Potkonjak. Power efficient organization of wireless sensor networks. *Proc. IEEE Int. Conf. Commun.*, 472–476, 2001.
33. M.B. Srivastava, A.P. Chandrakasan, and R.W. Broderson. Predictive system shutdown and other architectural techniques for energy efficient programmable computation. *IEEE Trans. VLSI Syst.*, 4, 42–55, 1996.
34. V. Swaminathan and K. Chakrabarty. Energy-conscious, deterministic I/O device scheduling in hard real-time systems. *IEEE Trans. Computer-Aided Design Integrated Circuits Syst.*, 22, 847–858, July 2003.
35. P.K. Varshney. *Distributed Detection and Data Fusion*, Springer, New York, 1996.
36. A. Wang, W.B. Heinzelman and A.P. Chandrakasan. An energy-efficient system partitioning for distributed wireless sensor networks. *Proc. IEEE Int. Conf. Acoustics, Speech, Signal Process.*, 2, 905–908, 2001.
37. Y. Zou and K. Chakrabarty. Sensor deployment and target localization based on virtual forces. *Proc. IEEE Infocom Conf.*, vol. II, session 1, paper 1, 2003.
38. Y. Zou and K. Chakrabarty. Energy-aware target localization in wireless sensor networks. *Proc. IEEE Int. Conf. Pervasive Computing Commun.*, 60–67, 2003.
39. Y. Zou and K. Chakrabarty. Target localization based on energy considerations in distributed sensor networks. *Ad Hoc Networks*, 1, 261–272, 2003.

9

Energy-Aware Routing and Data Funneling in Sensor Networks

Rahul C. Shah
University of California at Berkeley

Dragan Petrovic
University of California at Berkeley

Jan M. Rabaey
University of California at Berkeley

9.1 Introduction

The rising interest in ubiquitous sensor networks [12] has led to increased work on ad hoc multihop routing protocols. Unlike traditional routing protocols that minimize delay, many of these protocols try to minimize the energy required for communication because nodes in a sensor network are energy constrained. However, minimizing the energy consumption for every route can lead to undesirable effects like the creation of hotspots. These are areas that provide very good connectivity across the network and thus are used more often than other nodes. This leads to some nodes dying much earlier than others, resulting in lost sensing functionality as well as possible network partition. Many researchers have proposed ways to avoid this problem. This is typically done by using the residual energy at nodes as a routing metric rather than the energy used in communication.

Data communication in sensor networks primarily follows a pull model. In other words, most of the traffic in the network is based on a request–response model in which a request for information can set up a number of responses over a period of time. Thus, most routing protocols are reactive in nature, setting up route information during the request phase and using this information to route packets during the response phase. Based on this information, protocols use a single path or multiple paths to route the packets. Although single paths are cheaper to maintain, they run the risk of nondelivery of data if one node fails along the path. If such failure occurs, reflooding and route recomputation must be done to discover a new route.

Multipath protocols [13] get around this problem by using multiple routes so that failure of one route does not necessitate rediscovery of routes. However, using multiple routes increases the energy consumption for communication. Energy-aware routing [1] avoids this problem by discovering and using multiple routes; however, at any point in time, only one route is used. This is achieved by keeping a set of good routes and, for every packet, choosing a route in a probabilistic fashion. Thus, energy-aware routing is tolerant to failure of nodes (or, node mobility). It also uses the residual energy in its metric, thus avoiding nodes depleted of energy.

Energy-aware routing tries to optimize network lifetime — defined as the time until the first node in the network dies. Extending network lifetime translates to ensuring that energy use is equitable across the network. This is in contrast to simply minimizing the energy, which leaves the network with a wide disparity in the energy levels of the nodes and, eventually, disconnected subnets. If nodes in the network burn energy more equitably, then the nodes in the center of the network continue to provide connectivity longer, and the time to network partition increases. This leads to a more graceful degradation of the network.

Many different orthogonal techniques can be devised at the application or network level to extend the lifetime of the network. A first and obvious approach is to minimize the amount of data that must be transmitted through the network. One great opportunity arises from the observation that data transmitted by identical sensors spaced closely together tend to be spatially correlated. Thus, distributed compression schemes can be used to remove this redundancy and minimize the amount of traffic in the network [20].

A second approach to reduce the traffic volume (in terms of the number of bits an individual node must process) is to reduce transmission overhead. Packet payloads in sensor networks tend to be short (individual sensor measurements rarely need a resolution of more than 24 bits). Thus, packet headers comprise a substantial part of a packet because they contain training sequences for clock synchronization, source and destination information, and error control codes. These headers are necessary to make communication in a multihop network possible, but they do not carry any information about the phenomenon sensed. Thus substantial gains can be achieved if packets from different sources can be combined together and sent as one super-packet to the destination. This is the basic principle behind data funneling, a technique for routing with data aggregation [14]. Finally, energy-aware routing protocols can help to extend network lifetime compared to protocols that solely attempt to minimize network latency or the overall energy consumption of the network.

Section 9.2 of this chapter discusses the protocol stack design for sensor networks. Routing protocol characteristics and related work in this area are considered in Section 9.3. A linear programming formulation of routing is presented in Section 9.4. Section 9.5 introduces the energy-aware routing protocol, and Section 9.6 presents some simulation results. The data funneling protocol is described in Section 9.7 and the chapter concludes with Section 9.8.

9.2   Protocol Stack Design

This section gives a brief description of the application, network, and data link layers of the protocol stack.

9.2.1   Application Layer

Although the PicoRadio can support different types of applications, the driver application considered as an example in this chapter is environment control and monitoring. The aim is to control a typical office environment using a distributed building monitor and control approach. This is achieved by having three kinds of nodes in the system. The first are sensor nodes, which sense some environmental variable. The second node types are controllers, nodes that collect data from the sensors and, based on the data, decide the responses. They then command the third kind of nodes, actuators, to take appropriate action and affect the environment.

Based on the application, most of the nodes are expected to be static in nature, with a few low-speed mobile nodes. Furthermore, because the dominant form of data transport is from the sensors to the controller, it is important to optimize for that traffic. In addition, the total bit rate is rather low, about a few hundred bits per second per node. Also, because sensor data are inherently redundant, it is not necessary to have a transport layer that ensures reliable end-to-end delivery of every packet. If needed, the application can ensure the reliability itself. Thus, the application layer sends packets directly to the network layer.

9.2.2 Network Layer

The network layer has two primary functions: routing and addressing nodes. Although the rest of the chapter is concerned with routing, the kind of addressing used in the authors' network is briefly described in this section.

Traditional network addressing assigns fixed addresses to nodes, such as in the Internet. The advantage of such schemes is that the addresses can be made unique. However, a very high cost is associated with assigning and maintaining these kinds of addresses. This problem is exacerbated in mobile networks in which the topology information keeps changing. It is very difficult to route packets if the node address does not provide a clue as to the direction in which the packet is to be routed. Two approaches offer a solution to this problem. One is to maintain a central server that keeps up-to-date information on the position of every node. The other is to take the mobile IP approach: every node has a home agent that handles all the requests for the node and redirects them to the present position of the node.

For sensor networks, however, there is an important property of information flow that can be used advantageously. Most of the communication in sensor networks is of the form, "Give me the temperature of room 5." Thus, nodes can be addressed based on their geographical position. This information is also very useful for the routing protocol because it can direct communication in the right direction. Thus, for PicoRadio, class-based addressing is used. These addresses are triplets of the form (location, node type, node subtype). *Location* specifies a particular point or region in space that is of interest. *Node type* defines which type of node is required, such as sensor, controller, or actuator. Finally, the *node subtype* further narrows the scope of the address, such as temperature sensor, humidity sensor, etc. Thus, class-based addressing defines the type of node in the region of space that is needed.

In the rest of this chapter, class-based addressing within the network layer is assumed. Note that class-based addressing implicitly assumes that each node has knowledge of its position, which can be achieved by utilizing GPS or other distributed locationing algorithms [15]. These algorithms compute the location coordinates based on the received signal strength of neighboring nodes and the presence of certain *anchor* nodes in the network that know their exact positions. In this chapter, it is also assumed that all nodes know their position information perfectly.

9.2.3 Data Link Layer

The primary functions of the data link layer (DLL) are to provide access control, ID assignment, neighbor list management, and power control. The DLL coordinates assignments of local IDs so that each node gets a locally unique ID (up to two hops), while the IDs are reused globally. Thus a node can be identified within its neighborhood by its local ID. A common broadcast channel is used to send all DLL maintenance messages as well as for a rendezvous with another node when some data need to be sent. Once they rendezvous, the two nodes can communicate on other data channels. Using such a multichannel approach substantially reduces the number of collisions occurring during data transmission.

The link layer also keeps a list of its neighbors and metrics, such as the neighbor's position and the energy needed to reach it. This list is used by the network layer to make decisions regarding packet routing. Finally, the DLL also performs power control to ensure a power level that maintains an optimal number of neighbors.

9.3   Routing Protocol Characteristics and Related Work

Before introducing the energy-aware routing protocols, it is worthwhile to summarize the characteristics of what constitutes a good routing protocol in sensor networks. These are over and above properties traditionally required for routing protocols such as loop freedom and decentralized implementation. The additional requirements stem from the limited energy and small size of nodes in a sensor network.

- *Energy aware.* Sensor nodes may have limited battery lifetime and thus it is necessary for protocols to be aware of the residual energy at the nodes. Knowledge of the current energy use and the battery charge can enable the protocol to avoid nodes that are heavily depleted or to use energy-rich nodes.
- *Simple.* Simplicity of protocols refers to the fact that the protocol should not be too large or complex. It should minimize its memory requirement and also the amount of overhead it generates for routing. Thus, this refers to minimizing the communication and state of the protocol.
- *Adaptable.* Sensor nodes are inherently unreliable. They may fail due to lack of energy or may move to another region; thus, the routing protocol should be adaptable to such failures and be able to take appropriate action.
- *Scalable.* Because sensor networks can scale to hundreds and thousands of nodes, the routing must scale gracefully with such numbers. A key part of this is the size of the routing tables that are maintained and how they scale with the number of nodes/destinations in the network.

Now a small subset of routing protocols proposed for sensor and ad hoc networks will be discussed. The set is by no means exhaustive; however, it lists protocols that possess some of the preceding characteristics and from which energy-aware routing was derived.

Energy-aware routing is closely related to directed diffusion routing [2]. In directed diffusion, the destination sends an interest packet toward the source and sets up multiple routes in the process. The source subsequently sends data packets along the route, which has the minimum number of hops, consumes minimum energy, or something similar. Periodically, the source also sends data along all the paths to keep them alive and to check if any path has become better than the previously best route.

Both these protocols fall under the class of reactive protocols — protocols that discover routes only on demand. This is best exemplified by such ad hoc routing protocols such as ad hoc on-demand distance vector (AODV) routing [8] and dynamic source routing (DSR) [9]. AODV discovers routes on an as-needed basis by flooding the network and choosing the path with minimum delay. If the route gets broken, it discovers a new route by flooding the network again. On the other hand, DSR also discovers routes by flooding or by promiscuous listening to neighboring nodes to see what routes they have and then aggressively caching such information. It uses source routing, which means that the entire route is stored in the packet header rather than in intermediate nodes. Many reactive protocols also use the residual battery life as metrics for routing purposes [3, 4].

The other class of protocols is proactive protocols that maintain routes to all other nodes in the network. Destination sequenced distance vector (DSDV) routing [7] is one example of this kind. These protocols are not used often in sensor networks because they are better suited for high-traffic networks with a large number of source–destination pairs.

Location or geographic protocols make up one class of protocols very specific to ad hoc and, particularly, sensor networks. These protocols use the geographic position of nodes to forward packets toward the destination. This is very natural because nodes in a sensor network typically need to have a sense of their position to make sensor data meaningful. One of the first geographic routing protocols was developed in Karp and Kung [17], although there have been many others (e.g., [5, 18, 19]). Most of the work in this area until now has focused on routing around obstacles or areas in which no nodes are present.

Another class of routing protocols formulates the routing as a linear program (shown next) and tries to solve it in a decentralized fashion. The approach of Chang and Tassiulas [10] is representative of this kind of approach. Unlike most routing protocols which are based on heuristics, these protocols are based on a theoretical formulation of the problem. However, the problem is very difficult to solve in practice; to achieve a distributed implementation, it is necessary to use certain heuristics as well.

9.4 Routing for Maximizing Lifetime: A Linear Programming Formulation

As mentioned earlier, the routing protocol is designed to optimize network lifetime. Thus, it is very important to measure the performance of the protocol with respect to the optimum. The optimal protocol can be written as a linear program (as in Bhardwaj and Chandrakasan [16]). This linear program would correspond to a centralized routing scheme in which the traffic patterns are known *a priori*, in addition to the network topology and cost of communication between pairs of nodes. The cost of communication between pairs of nodes is the cost of communicating over the link by including the long-term average number of retransmissions needed for successful data delivery.

The metric to be optimized is the lifetime of the network (t), defined as the time when the first node in the network dies out. The first constraint is the non-negativity of the flow between any two nodes. Here, $r_{i,j}$ is the flow between nodes i and j. The second constraint is the flow constraint, which specifies that the amount flowing into a node is the same as the amount flowing out, except for the amount (S_i) that is absorbed or generated depending on whether the node is a sink or source, respectively. The third constraint is the energy constraint, which specifies that the energy expended in receiving and transmitting a packet should not exceed the initial amount of energy at the node. Solving this linear program would split each flow in such a fashion as to maximize the network lifetime.

$$\max t$$

$$s.t.$$

$$r_{i,j} \geq 0 \forall i, j$$

$$\sum_d r_{i,d} - \sum_s r_{s,i} = S(i) \cdot t$$

$$\sum_d P_{tx}(i,d)r_{i,d} + \sum_s P_{rx}(s,i)r_{s,i} \leq E_i \forall i$$

$$S(i) = \begin{cases} +1 & (source) \\ 0 & (relay) \\ -1 & (sink) \end{cases}$$

9.5 Energy-Aware Routing

The previous section specified optimal routing policy if centralized computation were possible; in reality the protocols need to work in a decentralized fashion. Thus, routes need to be selected based on some metric without full knowledge of the network. Even though sensor networks are energy limited, finding the lowest energy route and using that for every communication is not the best thing to do for network lifetime. The reason is that using a low-energy path frequently leads to energy depletion of the nodes along that path and, in the worst case, may lead to network partition.

To counteract this problem, a new protocol, called energy-aware routing, is proposed. The basic idea is that, to increase the survivability of networks, it may be necessary to use suboptimal paths occasionally. This ensures that the optimal path is not depleted and the network degrades gracefully as a whole rather than getting partitioned. To achieve this, multiple paths are found between source and destinations, and each path is assigned a probability of being chosen, depending on the energy metric. Every time data are to be sent from the source to destination, one of the paths is randomly chosen, depending on the probabilities. Therefore, none of the paths is used all the time, thus preventing energy depletion. Also, different paths are tried continuously, improving tolerance to nodes moving around the network.

Energy-aware routing is also a reactive routing protocol. It is a destination-initiated protocol in which the consumer of data initiates the route request and maintains the route subsequently. Thus, it is similar to diffusion in many ways. Multiple paths are maintained from source to destination. However, diffusion sends data along all the paths at regular intervals, but energy-aware routing uses only one path at all times. However, due to the probabilistic choice of routes, it can continuously evaluate different routes and choose the probabilities accordingly. The protocol has three phases:

- *Setup phase or interest propagation.* Directional flooding occurs to find all the routes from source to destination and their energy costs. This is when routing (interest) tables are built up.
- *Data propagation phase.* Data are sent from source to destination, using the information from the earlier phase. This is when paths are chosen probabilistically according to the energy costs calculated earlier.
- *Route maintenance.* Route maintenance is minimal. Directional flooding is performed infrequently from destination to source to keep all the paths alive and to collect new metrics.

9.5.1   Setup Phase

1. The destination node initiates the connection by flooding the network in the direction of the source node or region. It also sets the "cost" field to zero before sending the request.

$$Cost(N_D) = 0$$

2. Every intermediate node forwards the request only to neighbors closer to the source node (region) and farther away from the destination node than itself. Thus, at a node N_i, the request is sent only to a neighbor N_j that satisfies:

$$d(N_i, N_s) > d(N_j, N_s)$$

$$d(N_i, N_D) < d(N_j, N_D)$$

3. Upon receiving the request, the energy metric for the neighbor that sent the request is computed and is added to the total cost of the path. Thus, if the request is sent from node N_i to node N_j, N_j calculates the cost of the path as:

$$C_{N_j, N_i} = Cost(N_i) + Metric(N_j, N_i)$$

4. Paths that have a very high cost are discarded and not added to the forwarding table. Only the neighbors N_i with paths of low cost are added to the forwarding table FT_j of N_j.

$$FT_j = \left\{ i \,\middle|\, C_{N_j, N_i} \leq \kappa \cdot \left(\min_k C_{N_j, N_k} \right) \right\}$$

5. Node N_j assigns a probability to each of the neighbors N_i in the forwarding table FT_j, with the probability inversely proportional to the cost.

$$P_{N_j, N_i} = \frac{\dfrac{1}{C_{N_j, N_i}}}{\displaystyle\sum_{k \in FT_j} \dfrac{1}{C_{N_j, N_k}}}$$

6. Thus, each node N_j has a number of neighbors through which it can route packets to the destination. N_j then calculates the average cost of reaching the destination using the neighbors in the forwarding table.

$$Cost(N_j) = \sum_{i \in FT_j} P_{N_j, N_i} C_{N_j, N_i}$$

7. This average cost, $Cost(N_j)$ is set in the "cost" field of the request packet and forwarded along toward the source node, as in step 2.

9.5.2 Data Communication Phase

1. The source node sends the data packet to any of the neighbors in the forwarding table, with the probability of the neighbor being chosen equal to the probability in the forwarding table.
2. Each of the intermediate nodes forwards the data packet to a randomly chosen neighbor in its forwarding table, with the probability of the neighbor being chosen equal to the probability in the forwarding table.
3. This is continued until the data packet reaches the destination node.

The energy metric used in the protocol takes into account the residual energy of nodes along the path and the total energy needed for communication from the source to the destination. Thus, it was similar to the metric proposed by Chang and Tassiulas [10].

$$C_{ij} = e_{ij}^{\alpha} \cdot R_i^{-\beta}$$

Here, C_{ij} is the cost metric between two nodes i and j; e_{ij} is the energy used to transmit and receive on the link; and R_i is the residual energy at node i normalized to the initial energy at the node. The weighting factors α and β can be chosen to find the minimum energy path or the path with nodes having the most energy or a combination of these.

9.6 Simulations

Simulations were carried out in Opnet to demonstrate the increased network survivability due to energy-aware routing. The simulation consisted of 76 nodes in a typical office set up as in Figure 9.1. There were 65 sensors, seven static controllers, and four mobile nodes. Among the sensors, 47 were light sensors and 18 were temperature sensors. The controllers sent out requests for data to the sensors in their region of interest. These requests programmed the light sensors to send data every 10 sec and temperature data every 30 sec. These numbers are obtained from real application scenarios.

Every node consisted of an application and a network layer. The application layer was programmed to be a sensor or a controller, while the network layer performed the routing operations. Energy-aware routing was compared against directed diffusion routing. Both routing protocols used the same energy metrics for path selection; this was the metric function with $\alpha = 1$ and $\beta = 50$.

The MAC layer was abstracted away by providing for direct transfer of packets from the network layer of one node to the network layer of its neighbor. Thus, there was no contention for the medium when sending data. Fading effects were not considered either. Transmissions were always successful as long as a node was within the transmission range of the transmitter. The main purpose of removing the MAC was to orthogonalize the advantages of the network and media access layers and to evaluate the benefits of each separately.

Beyond its standard task of injecting, extracting, and forwarding packets, the network layer also maintained the neighbor list. An expanding ring search was used to create the list until it had the minimum number of neighbors (five) or the maximum radio range of the node was reached. Every node

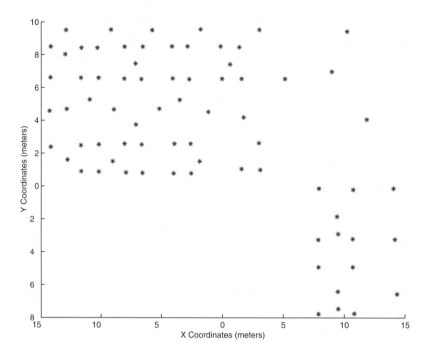

FIGURE 9.1 Layout of static nodes in the network.

was given an identical amount of initial energy at startup. Transmission used 20 nJ/bit + 1 pJ/bit/m^3 (i.e., energy drop off was r^3, which is a moderate indoor environment). The energy for reception was 30 nJ/ bit. These numbers are typical values for radios of the Bluetooth class. The packets were 256 b in size.

The performance of the protocols can also be checked against the optimal routing scheme using the linear programming formulation. This gives an idea of the maximum network lifetime if it were possible to use a centralized approach.

Figure 9.2 shows the results of one of the simulation runs. It shows the energy consumed by the various nodes during a 1-h period of the network using energy-aware routing. This can be compared against the energy consumed by the directed diffusion routing protocol in Figure 9.3. As expected, energy-aware routing spreads the traffic over the network, resulting in a much "cooler" network. As a consequence, the nodes in the center of the network conserve energy longer and the time until the first node runs out of energy increases.

The simulations show that energy-aware routing reduces the average energy consumption per node from 14.99 to 11.76 mJ, an improvement of 21.5% (Table 9.1). This is primarily due to the very low overhead of the protocol. At the same time, it reduces the energy differences between different nodes. Figure 9.4 shows energy consumption for the linear programming formulation. In such an optimal scenario, the controller nodes clearly are the bottleneck because they must process all the data packets traversing the network.

In another performance run, the network was simulated till a node ran out of energy. For diffusion routing, this occurred after 150 min; it took 216 min for the energy-aware routed network to fail. This is an increase in network lifetime of 44%, which agrees with the results of the previous simulation. In that simulation, the maximum energy use among all nodes was 57.44 mJ for diffusion and 41.11 mJ for energy-aware routing. This means that diffusion had a maximum energy consumption of 1.4 times energy-aware routing and thus an increase of 40% in the network lifetime is expected. Table 9.1 shows some of the statistics of energy consumption across nodes in the network for the three routing schemes.

It is worth noting that for linear program routing, the time to failure was 234 min, while the maximum energy use was 38.46 mJ, which was very close to energy-aware routing. Thus, this shows that energy-aware routing avoids heavily depleted nodes and performs very well compared to an optimal routing

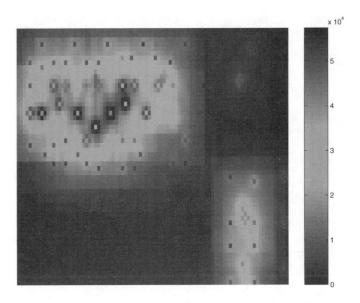

FIGURE 9.2 Energy consumption for energy-aware routing (μJ).

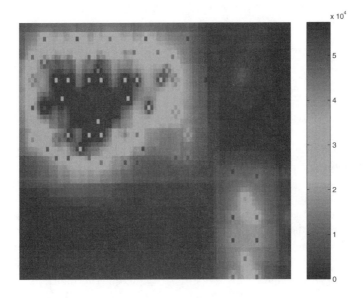

FIGURE 9.3 Energy consumption for diffusion routing (μJ).

TABLE 9.1 Energy Consumption Statistics after 1-h Simulation Time

Energy (mJ)	Avg.	Std. Dev.	Max	Min
Centralized	10.02	8.89	38.46	0.32
Diffusion	14.99	12.28	57.44	0.87
Energy-aware routing	11.76	9.67	41.11	0.98

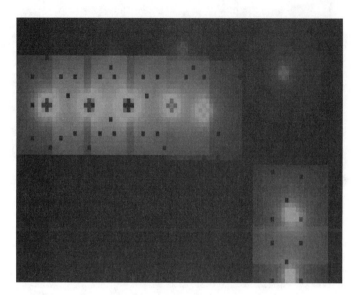

FIGURE 9.4   Energy consumption for linear program-based routing (μJ).

scheme. Also, the bit rate measured by the network is 250 b/sec, which demonstrates the extremely low data rate requirements of sensor networks. Thus, the results clearly show improved network health due to energy-aware routing.

9.7   Data Funneling

Although energy-aware routing tries to route data packets, avoiding regions in the network that are heavily depleted of energy, improvements in network lifetime are possible by reducing traffic volume in the network. To that end, this section discusses data funneling, a protocol that performs data aggregation while routing packets in the network. It is based on the energy-aware routing protocol, but instead of sending multiple packets from all nodes in a region, packets are aggregated along the way to obtain substantial energy savings.

The sensor networks envisioned by PicoRadio consist of a few controller nodes and many sensor nodes that periodically send their readings to the controllers. Because the controller nodes are required to have much greater computational and communication capabilities than the sensor nodes, the cost of controller nodes can be much greater than that of sensor nodes. Also, the controllers must decide what actions to take based upon collated readings from a large region of space. For these reasons, there are many more sensor nodes than controller nodes; each controller receives the readings of many sensors, while each sensor sends its data to only one or two controllers.

Furthermore, the amount of data in each reading is low, at most a few bytes of light, temperature, acoustic, seismic, or other measurements. However, packet headers include training sequences for clock synchronization; framing information; destination address; and error control codes and can be large relative to the packet size. Because many sensors report their data to the controller at approximately the same time and have similar headers, considerable savings can be realized by combining different packets into one large packet with a single header.

The main idea behind the algorithm, called data funneling, is that the controller breaks up the space into different regions (e.g., cuboids) and sends interest packets to each region, as shown in Figure 9.5. Upon receiving the interest packet, each node in the region will start periodically sending its readings back to the controller at an interval specified in the interest packet, usually every few minutes. Because many or all of the nodes within the region will be sending their readings to the controller at the same time, it would be much more efficient to combine these readings into a single packet so that only one

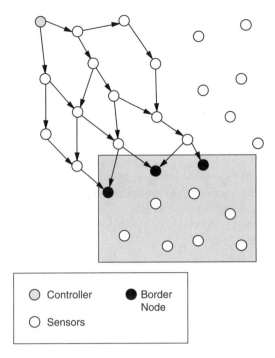

FIGURE 9.5 Control node, target region, directional flood, and border nodes in data funneling.

packet with only one header travels from the region to the controller. The question is how all these readings can be collected at a single point and combined into a single packet.

The data funneling algorithm works as follows. The interest packets are sent toward the region using directional flooding. Each node that receives the interest packet checks whether it is in the target region. If it is not, it computes its cost for communicating back to the controller, updates the cost field within the interest packet, and sends it on toward the specified region. This is the directional flooding phase.

When a node in the target region receives the interest packet from a neighbor node outside the target region, the directional flooding phase concludes. The node realizes that it is on the border of the region and designates itself to be a border node as shown in Figure 9.5. Each border node computes its cost for communicating with the controller in the same manner as was done by the nodes outside the region during the directional flooding phase. It then floods the entire region with a modified version of the interest packet. The "cost to reach the controller" field is reset to zero and becomes the "cost to reach the border node field." Within the region, each node only keeps track of its cost for communicating with the border node, not its cost for communicating with the controller. Intuitively, it is as if the border node becomes the controller of the specified region. At one of the border nodes, all the readings from within the region will be collated into a single packet.

In addition, two new fields are added to the modified interest packet. One keeps track of the number of hops that have been traversed between the border node and the node currently processing the packet. The other field specifies the border node's cost for communicating with the controller; this field, once defined by the border node, does not change as the packet travels from one node to another.

Once the nodes within the region receive the modified interest packet from the border nodes, they will then route their readings to the controller via each of the border nodes in turn. Several border nodes are within the region, so maximizing aggregation of sensor readings requires all the nodes within the region to agree to route their data via the same border node during every given round of reporting back to the controller. This is accomplished by having every node compute an identical schedule of which border node to use during each round of reporting. This is achieved by each node in the region applying the same deterministic function to the vector of costs to reach the controller seen by each border node.

Because all the nodes apply the same function to the same inputs, they will all compute the same schedule, allowing them to collect all of their data at one border node during each round of reporting. The function used to compute the schedule can be the function used to compute the probabilities for selecting different paths in energy-aware routing. This allows border nodes with a low cost for communicating to the controller to be used more frequently than the ones with a high cost.

As data flow within the region from the sensors to the border nodes, the packets can be aggregated along the way as shown in Figure 9.6. When the time comes to send a new round of observations back to the controller, the sensor nodes do not immediately start sending their packets. Instead, before sending their readings toward the border node to be used in that round of reporting, they wait an amount of time inversely proportional to their distance (in number of hops) to that border node. This allows nodes far away from the border node to send their data earlier than nodes closer to the border node. Thus, nodes close to the border will first receive the readings from upstream nodes and bundle those readings with their own. In the end, all of the data to be sent out by all the nodes within the region will be collated at one border node and sent back to the controller in a single packet, as shown in Figure 9.6. The protocol is summarized next.

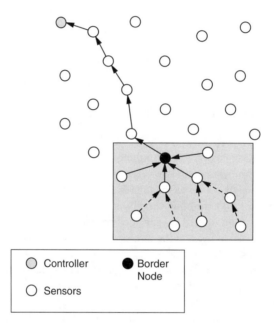

FIGURE 9.6 Funneling data within the region and reporting back to the controller.

9.7.1   Setup Phase

1. The controller divides the area it wishes to monitor into cuboids.
2. It then initiates a directional flood toward each region.
3. Each intermediate node records the cost of reaching the controller as in the routing scheme described previously.
4. When the packet reaches the region, the first node in the region that receives the packet designates itself as a border node.
5. The border node adds two new fields to the packet: the cost for reaching the border node and the number of hops to the border node. It then floods the region with this modified packet.
6. All nodes within the region receive packets from all the border nodes. Based on the energy required by each border node to reach the controller, they compute a schedule of border nodes at which data are to be aggregated.

9.7.2   Data Communication Phase

1. When a sensor has a data sample that it needs to send back to the controller, it uses the schedule to figure out the border node to use in the current round of reporting.
2. It then waits for a time inversely proportional to the number of hops from the border node before sending out the packet.
3. Along the way to the border node, the data packets are joined together until they reach the border node.
4. The border node collects all the packets and then sends one packet with all the data back to the controller, using probabilistic routing.

Data funneling creates clusters within the sensor network, but does so in a fluid fashion, which makes the approach a lot less brittle. There is no single cluster head whose failure can be devastating to the functionality of the network. Instead, the border nodes take turns acting as the cluster head, spreading out the responsibility and the load (i.e., energy consumption) among them. Also, the controller can redefine the regions into which its area of interest is divided, thereby forcing the nodes to divide into new clusters and elect new sets of border nodes. The controller can redefine the regions based on the data received from the nodes and/or the energy remaining in the nodes so as to ensure that nodes with the greatest energy reserves act as border nodes.

To demonstrate its feasibility, the data funneling algorithm was implemented in the Opnet network simulator. The simulation was performed only for the network layer; lower layers were abstracted away. Energy consumed at all the nodes for transmission, reception, and computation was measured. One sample topology is shown in Figure 9.7. The controller node, shown within the dark circle, queried a region, shown as the large rectangle, containing 15 sensor nodes. Copies of the interest packet propagated toward the region, and the four nodes shown within the squares were determined to be the border nodes. Each of the sensors sent its readings to the controller every 10 sec, and the packets were aggregated along the way. The simulation measured the number of sensor readings contained within each transmitted packet.

For the topology shown in Figure 9.7, the average number of sensor readings per transmitted packet was seven. This means that the energy expended by the network on transmitting packet headers was reduced by 86%. In general, the larger the region and the further away it is from the controller, the greater the savings are due to funneling. If n sensors are in the region and the region is far away from the controller, the energy spent on transmitting headers will be reduced by a factor of approximately n.

Let γ be the ratio of bits in a packet header to the total number of bits in a packet containing the header and a single sensor reading for a particular application; let m be the average number of sensor readings per transmitted packet when data funneling is employed. Then, the total energy expended by the network on communication is reduced by $\gamma \cdot \dfrac{m-1}{m} \cdot 100$ % due to data funneling if no compression of the sensor readings is done at the aggregation points. Performing compression on the sensor readings at the aggregation points within a region would result in even greater energy savings.

9.8   Conclusion

This chapter discussed the concept of the lifetime of sensor networks and presented two approaches to extend it. The first protocol, energy-aware routing, uses a very simple energy metric based on communication energy and residual battery lifetime at the nodes to figure out a set of good paths. These paths are then chosen probabilistically to route packets so that no single path is depleted. This leads to more uniform depletion of nodes across the network. Currently, the choice of the energy metric is based on heuristics and previous research; further research into the energy metric used may lead to better results.

However, upon implementing the protocol on a real-life test bed, it was found that, although the network resources depleted uniformly as expected, in some situations the link with a particular node

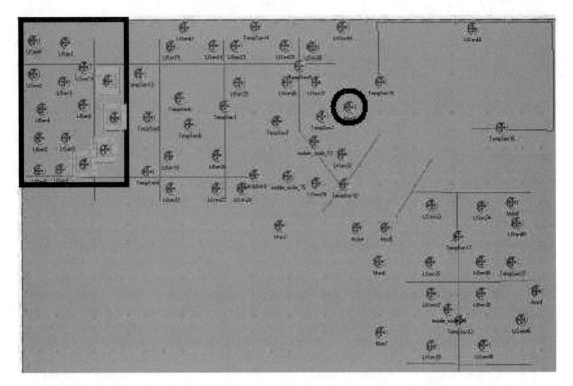

FIGURE 9.7 Sample topology for data funneling.

was bad. If, unfortunately, the node selected that link as the next hop, it would take a while for the transmission to be successful. Although this is inevitable in a wireless scenario, it still might be possible to exploit the fact that other nodes may have a good link at the same point in time. In other words, using current knowledge of the link state with nodes, either as part of the energy metric or otherwise, might lead to improved performance.

The second protocol presented uses energy-aware routing as the primary routing mechanism while performing aggregation of data. Combining data from a set of sensors in the same area can lead to substantial savings as the number of packet headers reduce. Data funneling achieves this in a completely distributed fashion and without the need for any local coordination. The algorithm may also be used in conjunction with other routing protocols and can be combined with other source-coding techniques to reduce network energy consumption further.

Acknowledgments

This work was supported by DARPA as a part of the PAC/C program and NSF as a part of the CITRIS project. Their support is greatly appreciated. The support of the Berkeley Wireless Research Center member companies is kindly appreciated also.

References

1. Shah R.C. and Rabaey J.M., Energy-aware routing for low energy ad hoc sensor networks, *IEEE WCNC*, 1, 350–355, 2002.
2. Intanagonwiwat C., Govindan R., and Estrin D., Directed diffusion: a scalable and robust communication paradigm for sensor networks, *IEEE/ACM Mobicom*, 2000, 56–67.
3. Singh S., Woo M., and Raghavendra C.S., Power-aware routing in mobile ad hoc networks, *IEEE/ACM Mobicom*, 1998, 181–190.

4. Toh C.K., Maximum battery life routing to support ubiquitous mobile computing in wireless ad hoc networks, *IEEE Commun. Mag.*, June 2001, 138–147.

5. Jain R., Puri A., and Sengupta R., Geographical routing for wireless ad hoc networks using partial information, *IEEE Personal Commun. Mag.*, Feb. 2001, 48–57.

6. Royer E. and Toh C.K., A review of current routing protocols for ad hoc mobile wireless networks, *IEEE Personal Commun. Mag.*, April 1999.

7. Perkins C.E. and Bhagwat P., Highly dynamic destination sequenced distance vector routing (DSDV) for mobile computers, *Comp. Commun. Rev.*, Oct. 1994, 234–244.

8. Perkins C.E. and Royer E., Ad hoc on demand distance vector routing, *Proc. 2nd IEEE Workshop Mobile Comp. Syst. Apps.*, Feb. 1999, 90–100.

9. Johnson D.B. and Maltz D.A., Dynamic source routing in ad hoc wireless networks, in *Mobile Computing*, Kluwer, 1996, 153–181.

10. Chang J. and Tassiulas L., Energy conserving routing in wireless ad hoc networks, *IEEE Infocom*, 2000, 22–31.

11. Rabaey J.M. et al., PicoRadio supports ad hoc ultra-low power wireless networking, *IEEE Computer Mag.*, July 2000, 42–48.

12. Akyildiz I., Wireless sensor networks: a survey, *Computer Networks*, 38(4), 2002, 393–422.

13. De S., Qiao C., and Wu H., Meshed multipath routing: an efficient strategy in sensor networks, *IEEE WCNC*, 2003.

14. Petrovic D., Shah R.C., Ramchandran K., and Rabaey J.M., Data funneling: routing with aggregation and compression for wireless sensor networks, *IEEE SNPA*, 2003.

15. Savarese C., Langendoen K., and Rabaey J.M., Robust positioning algorithms for distributed ad hoc wireless sensor networks, *Proc. 2002 USENIX Annu. Tech. Conf.*

16. Bhardwaj M. and Chandrakasan A., Bounding the lifetime of sensor networks via optimal role assignments, *IEEE Infocom*, 2002, 1587–1596.

17. Karp B. and Kung H.T., GPSR: greedy perimeter stateless routing for wireless networks, *IEEE/ACM Mobicom*, 2000, 243–254.

18. Stojmenovic I., Position-based routing in ad hoc networks, *IEEE Commun. Mag.*, 40(7), 128–134, July 2002.

19. Ko Y. and Vaidya N., Location-aided routing (LAR) in mobile ad hoc networks, *IEEE/ACM Mobicom*, 1998, 66–75.

20. Chou J., Petrovic D., and Ramchandran K., A distributed and adaptive signal processing approach to reducing energy consumption in sensor networks, *IEEE Infocom*, 2003.

10

Reliable Energy-Constrained Routing in Sensor Networks

Rajgopal Kannan
Louisiana State University

Lydia Ray
Louisiana State University

S. Sitharama Iyengar
Louisiana State University

Ram Kalidindi
Louisiana State University

10.1 Introduction

A wireless sensor network is an autonomous system of numerous tiny sensor nodes equipped with integrated sensing and data processing capabilities [1]. Sensor networks are distinguished from other wireless networks by the fundamental constraints under which they operate: (1) sensor nodes are untethered; and (2) sensor nodes are unattended. These constraints imply that network lifetime, i.e., the time during which the network can accomplish its tasks, is finite. Therefore, sensors must utilize their limited and unreplenishable energy as efficiently as possible.

The energy efficiency of routes is an important parameter; however, maximizing network information utility and lifetime implies that the *reliability* of a data transfer path from reporting to querying sensor is also a critical metric. This is especially true given the susceptibility of sensor nodes to denial-of-service (DoS) attacks and intrusion by adversaries that can destroy or steal node data [11]. The possibility of sensor node failure due to operation in hazardous environments cannot be discounted, especially for environmental monitoring and battlefield sensor network applications. For such networks to carry out their tasks meaningfully, sensors must route strategic and time-critical information via the most reliable paths available. Thus, an additional constraint on sensor operations can be introduced: sensor s_i can fail with probability $q_i = 1 - p_i$.

The primary issue addressed in this chapter is reliable energy-constrained intercluster routing within the framework of hierarchical cluster-based sensornet architectures. In a hierarchical architecture [3], nodes in close proximity form clusters, with one node in each cluster designated or elected as the cluster head with special responsibilities. Traffic between different clusters is routed through their corresponding cluster heads. Most hierarchical architectures are based on the assumption that cluster heads can communicate directly with each other. Here, discussion concerns a more realistic two-level hierarchical architecture in which cluster heads called *leader nodes* must use the underlying network infrastructure for communication, i.e., leader–leader and leader–sink routing.

Network partition is expedited by uneven energy distribution across sensors, resulting from improperly chosen routes. Ideally, data should be routed over a path in which participating nodes have higher energy levels relative to other nonparticipating nodes. Network operability will be prolonged if a critically energy-deficient node can survive longer by abstaining from a route rather than taking part in a route for a small gain in overall latency. Similarly, routing over less reliable paths increases energy depletion due to retransmissions. Therefore, *path length*, *path reliability*, and *path energy cost* are critical metrics affecting sensor lifetime.

Recent research in the literature has begun to consider these aspects. For example, Shah and Rabaey [4] describe a probabilistic routing protocol in which non least-energy cost paths are chosen periodically. In Yu et al. [5], a node attempts to balance energy across all its neighbors while finding shortest paths to the sink. However, no unified analytical model explicitly considers routing under the constraints of energy efficiency, path length, and path reliability. The choices of sensor nodes under these constraints are a natural fit for a game-theoretic framework.

This chapter describes a game-theoretic paradigm for solving the problem of finding reliable energy-optimal routing paths with bounded path length and defines two routing games in which sensors obtain benefits by linking to healthy and reliable nodes while paying a portion of path length costs. Thus, sensor nodes modeled as intelligent agents cooperate to find optimal routes. This model has the following benefits:

- Each sensor will tend to link to more reliable and healthier nodes; thus, network partition will be delayed.
- Because each node shares the path length cost, path lengths will tend to be as small as possible; therefore, delay is restricted in this model. Also, shorter path lengths will prevent too many nodes from taking part in a route, thus reducing overall energy consumption.

The Nash equilibria of these routing games define optimal reliable length energy-constrained paths. Computing optimal paths is NP-hard in arbitrary sensor networks, but it can be found in polynomial time (in a distributed manner) in sensor networks operating under a geographic routing regime. The following sections describe fully distributed, scalable, nearly stateless and easily implementable protocols for reliable and length energy-constrained intercluster routing.

10.2　Game-Theoretic Models of Reliable and Length Energy-Constrained Routing

Let $S = \{s_1, s_2, ..., s_n\}$ be the set of sensors in the sensor network participating in the routing game. Let $s_r = L_1$ and $s_q = L_2$ be a pair of leader nodes. Data packets are to be routed from L_1 to L_2 through an optimally chosen set, $S' \subset S$, of intermediate nodes by forming communication links.* Let $v_i \geq 0$ denote the value of information at sensor s_i to be routed to L_2, with $q_i = 1 - p_i$ the probability of sensor failure. Note that multicast communication between sets of leader nodes is not considered.

Strategies. Each node's strategy is a binary vector $I_i = (I_{i1}, I_{i2}, ..., I_{ii-1}, I_{ii+1}, ..., I_{in})$, where $I_{ij} = 1$ ($I_{ij} = 0$) represents sensor s_i's choice of sending/not sending a data packet to sensor s_j. Because a sensor typically relays a received data packet to one neighbor only, it is assumed that a node forms only one link for a given source and destination pair of leader nodes. In general, a sensor node can be modeled as having a mixed strategy [10], i.e., the I_{ij}'s are chosen from some probability distribution. However, in this chapter, the strategy space of sensors is restricted to pure strategies only. Furthermore, in order to eliminate some trivial equilibria, each sensor's strategy is nonempty and strategies resulting in a node linking to its

---

*In general, sensors in S will be simultaneously participating in routing paths between several such pairs.

ancestors (i.e., routing loops) are disallowed. Consequently, the strategy space of each sensor s_i is such that prob. $[l_{ij} = 1] = 1$ for exactly one sensor s_j and prob. $[l_{ij} = 1] = 0$ for all other sensors, such that no routing loops are formed [7].

Payoffs. Let $l = l_1 \times l_2 \times \ldots \times l_n$ be a strategy in the routing game resulting in a route P from source to destination leader node. Each sensor on P derives a payoff from participating in this route.

Reliable query routing (RQR) payoff model. Every sensor that receives data has an incentive for reaching the destination leader node s_q; thus, the benefit to any sensor s_i on P must be a function of the path reliability from s_i onwards. Because the network is unreliable, the benefit to player s_i should also be a function of the expected value of information arriving at s_i. Therefore, the payoff at s_i on linking to node s_j in P can be written as:

$$\prod_i(l) = \begin{cases} V_i R_i - c_{ij} & \text{if } s_i \in P \\ 0 & \text{otherwise} \end{cases}$$

where R_i denotes the path reliability from si onwards to s_q and $V_i = v_i + p_k V_k$ denotes the expected value of the data at node i with parent s_k in P.

Length energy-constrained routing (LEC) payoff model. In addition to reliability, leader–leader routing protocols must be designed to dissipate energy equitably over sensors. One possible approach is to prevent low-energy nodes from taking part in a route as long as they are energy-deficient relative to their neighbors. However, a route that focuses only on energy efficiency may be undesirably long because the lowest energy-cost path need not be the shortest. Conversely, longer paths will result in energy depletion at more sensors while also increasing delay.

Under this model, the payoff of sensor s_i on linking to s_j in P is defined as:

$$\pi_i(l) = E_j - \xi L(P) \tag{10.1}$$

where E_j is the residual energy level of node s_j and $L(P)$ the length of routing path P. E_j represents a benefit to s_i, thus inducing it to forward data packets to higher energy neighbors. The parameter ξ represents the proportion of path length costs borne by sensor s_i. Choosing ξ as a positive constant or proportional to path length will inhibit formation of longer routing paths. Conversely, setting ξ at zero or inversely proportional to path lengths will favor the formation of paths through high-energy nodes. Zeta is chosen as a nonzero positive constant for this routing game; thus, each sensor will forward packets to its maximal energy neighbor in such a way that the length of the path formed is bounded. This model encapsulates the process of decentralized route formation by making sensor nodes cooperate to achieve a joint goal (shorter routing paths) while optimizing their individual benefits.

A Nash equilibrium of this game under both payoff models corresponds to the path in which all participating sensors have chosen their best-response strategy, i.e., the one that yields the highest possible payoff given the strategies of other nodes. This equilibrium is the optimal reliable energy-constrained (RQR) and optimal length energy-constrained (LEC) path in the sensor network for the given leader pair. Note that the process of determining the optimal path requires each node to determine the optimal paths formed by each of its possible successors on receiving its data. The node then selects as next neighbor that node, the optimal path through which gives the highest payoff.

10.2.1 Reliable Routing in Geographically-Routed Sensor Networks

Consider the reliable routing problem for sensor networks in which sensors are restricted to following a *geographic routing* regime. In other words, the strategy space of each sensor in the RQR game includes only neighbors geographically closer to the destination than it is. Routing paths under this regime are implicitly length constrained. For each sensor, it is assumed that the set of downstream neighbor nodes to a given destination can be found using a global positioning system (GPS) or some other localization protocol.

Let G be an arbitrary sensor network following geographic routing, with sensor success probabilities P, communication energy costs C, and data of value v, to be routed from leader node s_r to the sink/leader node s_q, where $v_i = 0 \forall i \neq r$. Although the RQR problem is *NP*-hard for general sensor networks, it becomes surprisingly easy when the additional constraint of path length [7] is added.

Lemma 10.1. Let L_i be the longest geographically-routed path from s_i to s_q in G. Then, s_i can determine its optimal RQR neighbor under the reliability payoff model in $|L_i|$ steps.

PROOF. The following simple observation is noted first: in a geographically-routed network, all feasible routing paths from s_r to any node s_i and from s_i to the sink s_q intersect only at s_i. If any other such node existed it would need to be geographically closer than s_i to s_r (because it is on a feasible path from s_r to s_i) as well as to s_q (because it is on a feasible path from s_i to s_q), which is impossible.

Let $R(P_i(v_i))$ represent the reliability of the optimal RQR path P_i from s_j to s_q, transmitting information of value v_i. From the preceding observation, s_{ij} merely needs to know optimal values to s_q from each of its downstream neighbors. Let D_i represent this set; then the optimal neighbor for s_i is

$$N_{opt}(v_i) = \underset{s_j \in D_i}{\arg\max} \left\{ v_i p_i R(P_j(p_i v_i)) - c_{ij} \right\} \tag{10.2}$$

where v_i is the expected value of information received at s_i from a given upstream neighbor. The number of such values is proportional to the number of paths from s_r to s_i, which can be exponentially large. However, these values can be divided into disjoint, contiguous intervals in $(0..v_r]$, which makes next-hop selection much easier.

The lemma can now be formally proved by induction. Consider node s_i, whose longest path to the destination is of length one. It will link directly to s_q for all values vi; $p_i p_q v_i > c_{iq}$. s_q is unreachable for smaller values of vi. Thus, at node s_j, the optimal choices are divided into tuples consisting of (two) value intervals and optimal path reliabilities corresponding to each interval.

During the kth step of the algorithm, all nodes with $|L_i| = k$ follow the same reasoning, based on the optimal choices of downstream nodes in step $k - 1$. Each node has multiple optimal neighbors, based on a division of the incoming information value into disjoint intervals in $(0..v_r]$. These intervals are polynomial in number and calculated at each node on the basis of intersections of value intervals and optimal reliabilities from its downstream neighbors.

10.2.2 Distributed Implementation of Length-Constrained RQR

The results presented Kannan et al. [7] are now summarized. Let $Di = \{s_{i_1}, s_{i_2}, ..., s_{i_l}\}$ be the set of downstream next-hop neighbors of s_j. For each node s_{ij} in this set, let the expected values of incoming information be divided into N_{ij} disjoint consecutive intervals $I_1^{i_j}, I_2^{i_j}, ..., I_{N_{ij}}^{i_j}$, where $U_t I_t^{i_j} = (0, v_r]$. Let $B(I_t^{i_j})$ and $E(I_t^{i_j})$ denote the (open) left and (closed) right endpoints and let $R(I_t^{i_j})$ be the optimal path reliability from s_{ij} onwards for information of expected value in the given interval $I_t^{i_j}$. When information of expected value v_i arrives at s_i and is forwarded, the expected value of information at s_{ij} is $p_i v_i$. Therefore, each value interval at s_{ij} corresponds to an equivalent "stretched" interval at s_i with left endpoint $B()/p_i$ and right endpoint $\max(v_r, E()/p_i)$. Henceforth, the notation I_t^{ij} refers to the stretched interval at si rather than the actual interval at s_{ij}.

Let $\pi_i(i_j, v_i, I_t^{i_j})$ represent the payoff to sensor si on sending information of value $v_i \in I_t^{i_j}$ to downstream neighbor s_{ij}. Note that the payoff function is continous and increasing through the entire range of v_i (as v_i increases, the payoff can only increase). It can therefore be assumed that all intervals give a positive payoff because intervals with negative or zero payoff can be identified and removed. The following lemma shows that the payoff optimality of two intersecting intervals at different neighbors s_{ij} and s_{ik} can be determined using a single fixed point.

Lemma 10.2. If $\pi_i(i_j, v_i, I_t^{i_j}) < \pi_i(i_k, v_i, I_u^{i_k})$ for $v_i = \text{Inf}[I_t^{i_j} \bigcap I_u^{i_k}]$, then $\pi_i(i_j, v_i, I_t^{i_j}) < \pi_i(i_k, v_i, I_u^{i_k})$ for all

$v_i \in [I_t^{i_j} \bigcap I_u^{i_k}]$. *If the two payoffs are equal at the fixed point, then $\pi_i(i_j, v_i, I_t^{i_j}) \leq \pi_i(i_k, v_i, I_u^{i_k})$ throughout the*

intersection if $R(I_t^{i_j}) \leq R(I_u^{i_k})$.

The lemma follows by definition of the payoff function in Equation 10.2. Thus, to compare two different intervals, it is necessary only to evaluate their payoff at the smallest intersecting point. Lemma 10.2 can be used to compute value ranges and corresponding optimal next neighbors (i.e., those that maximize payoff in the given value range) at each node, provided the optimal solutions are available at nodes one hop away. This can be achieved using reverse directional flooding of control packets [7] from the sink to the source.

Theorem 10.1. The optimal length-constrained RQR path in a sensor network with geographic routing can be computed in a distributed manner using reverse directional flooding with $O(|E|)$ total messages, where E is the number of edges in the sensor network. Optimal neighbors at each node can be found in $O(N_T log|D_i|)$ time [7].

10.3 Distributed Length Energy-Constrained (LEC) Routing Protocol

A distributed implementation of the LEC routing protocol is described in terms of a simplified "team" version of the routing game. The protocol is derived from work presented in *Lecture Notes in Computer Science* [8]. This team-game routing protocol can be easily modified to obtain optimal LEC paths as well. In the team LEC game, each node on a path shares the payoff of the worst-off node on it. Formally, let L be the set of all distinct paths from a particular source and destination leader pair. Let $E_{min}(P)$ be the smallest residual energy value on path P. Then the equilibrium path of the team LEC game is defined as:

$$\hat{P} = \text{argmax}_{P \in L} (E_{min}(P) - \xi |P|) \tag{10.3}$$

For simplicity in the protocol description later, ξ is set to zero. However, the protocol can be easily modified for nonzero ξ as well as for computing optimal LEC paths in the original LEC game. The optimal path under this condition is interpreted as follows: given any path P, the durability of the path is inversely proportional to $E_{min}(P)$. A path with lower average energy but higher minimum energy should last longer than a route with the opposite attributes because the least energy node is the first to terminate and make that route obsolete. Thus, the inverse of the minimum node energy on a given path reflects the energy weakness of the path. The proposed protocol will select an optimal path of bounded length with the least energy weakness. Node energy levels are changing continuously in a sensor network due to sensing, processing, and routing operations, so the optimal path needs to be recomputed periodically. Therefore, the proposed protocol operates in two different phases: data transmission and path determination as described next.

10.3.1 Data Transmission Phase

During this phase, data packets are transmitted from one leader node to the other through the optimal path (with least energy weakness). Each data packet also potentially collects information about the energy consumption en route by keeping track of residual energy levels of nodes on the path. When energy levels of a given critical number of nodes fall below a certain threshold, the data transmission phase ends and the new optimal path determination phase begins. The fundamental steps of the data transmission phase are:

- Each data packet is marked by the source leader node with the geographical position of the destination node and with a threshold value *th*; each packet contains a special *n*-bit energy depletion indicator (EDI) field, where $n \ll$ packet size.
- Each sensor node receiving a data packet determines whether its energy level has fallen below the threshold *th*. If so and if the EDI field in the data packet is not exhausted, the node sets a single bit in the EDI field. Then it forwards the packet to the best next-hop neighbor according to its routing table. It is assumed that before the network starts any activity, all ordinary nonleader sensor nodes have the same energy level. Therefore, during the first data transmission phase, the best next-hop neighbor of a node is the one geographically nearest to the destination leader node. In all other phases, the routing table is updated according to the optimal LEC path calculation.
- If the receiver leader node gets a data packet with all *n* bits in the EDI field set to 1, it triggers a new optimal path selection procedure.

10.3.1.1 Calculation of the Threshold Value

The threshold value *th* plays a very important role in the data transmission phase because it is used to provide an approximate indication that the current optimal path has become obsolete. Intuitively, *th* must be a function of the current residual node energy levels in the network. In this chapter, the following function is used:

$$th = \beta E_{\min} \tag{10.4}$$

where $0 < \beta < 1$ and $E_{\min}$ is the minimum energy level in the current optimal path. Because $E_{\min}$ changes with time, the threshold is recalculated in each path determination phase, consistent with current energy distribution across the network.

10.3.2 Path Determination Phase

This phase begins when the destination leader node receives critical EDI information and ends when the sending leader node has updated its routing table and recalculated the threshold value. The principal steps are:

- The destination leader node L_2 triggers this phase by flooding the network with control packets along the geographic direction of the source leader node L_1. Note that this *reverse directional flooding* occurs in a direction opposite that of data transfer.
- Each node forwards control packet to all its neighbors in the geographic direction of L_1. Each control packet contains a field EM_p that indicates the maximum of the minimum energy levels of all partial paths converging at the given node, i.e., the inverse of the energy weakness of the strongest partial path.
- On receiving the first control packet, each node sets a timer for a prefixed interval *T*. This time period should be large enough for the node to receive future control packets from most of its neighbors (corresponding to different partial paths from the leader node terminating at this node), but not large enough to cause high delays. With each arriving control packet, the node updates and stores the highest EM_p value seen so far. However, if its own energy level E_i is lower than all these values, it stores E_i. With each control packet, it also updates its routing table for destination L_2 to point to the node from which it has received the highest energy control packet. Note that this part of the protocol can be easily modified to incorporate path lengths in addition to the preceding minimum energy computations.
- When the timer expires, this node forwards a new control packet with EM_p field set to the stored energy value to all its neighbors in the geographic direction of L_1. Control packets arriving after the timer expires are discarded.
- Eventually, L_1 begins receiving control packets and sets its timer. Its value of *T* can be determined in many ways, depending on the specific requirements of applications. In this chapter, *T* is

calculated to ensure that most of the paths from L_1 to L_2 are included in the optimality calculations. If (D_{max}) is the maximum transmission delay between two nodes, the value of T is determined as $(MINHOP * D_{max})$, where *MINHOP* is an estimate of the shortest path from L_1 to L_2. This value can be estimated *a priori* using GPSR routing [6] before the first data transmission phase. Note that the given value of T allows control packets from paths up to twice the length of the shortest path to be forwarded to $L - 1$. Also note that D_{max} is a function of the specific MAC-layer protocol implemented in the sensor network. Finally, when the timer expires at L_1, it selects the final E_{min} value as the highest EM_p value received, calculates the new value of *th*, and sets its routing table accordingly. The next data transmission phase can now begin.

10.3.3 Selection of β

Data transmission in the proposed protocol ends when residual energy levels of at least n nodes on the current path fall below threshold *th*. With high β, the smaller the threshold value is, the larger the useful data transmission phase. If the traffic is fairly bursty, *th* should be large so that data packets in the burst can be transmitted. Beta is an empirical value and can be modified based on previous observations. A useful rule of thumb to set the value of β for the current period is as follows:

$$\beta_{current} = \alpha\beta_{prev1} + (1-\alpha)\beta_{prev2} \tag{10.5}$$

where $0 \le \alpha \le 1$, β_{prev1} and β_{prev2} and are the previous and previous-to-previous values of β, respectively. Alpha should be chosen according to the specific requirement, i.e., whether the current value of β should be increased or decreased and to what extent.

10.3.4 Selection of Energy Depletion Indicator

Energy depletion indicator is an integer that indicates the maximum number of critical nodes allowed during a data transmission period. The main contribution of an energy depletion indicator is to regulate the duration of the data transmission phase. The higher the value of this parameter is, the longer the period of data transmission. Like β, this parameter is also empirical and can be modified based on previous observations. A rule of thumb similar to that of β can be used to modify the value of this parameter. Thus,

$$EDI_{current} = \gamma EDI_{prev1} + (1-\gamma)EDI_{prev2} \tag{10.6}$$

where $0 \le \gamma \le 1$, EDI_{prev1} and EDI_{prev2} are the previous and previous-to-previous values of *EDI*, respectively. Gamma should be chosen according to the specific requirement, i.e., whether the duration of the data transmission phase should be increased or decreased and to what extent.

10.4 Performance Evaluation

The main objective of the protocol is gradually to balance energy consumption across the network. To evaluate performance of this protocol, the following metrics, which reflect dispersion or concentration of energy consumption across a network, are used.

- *Variance of energy level.* The variance of the energy levels of all the nodes is the primary measure of dispersion. A high variance indicates higher energy consumption at some of the nodes compared to others.
- *Range of energy level.* This metric measures the difference between the energy levels of the maximum energy node and the minimum energy node over the entire network. A large value for this range is a result of unfair distribution of routing load among the nodes.

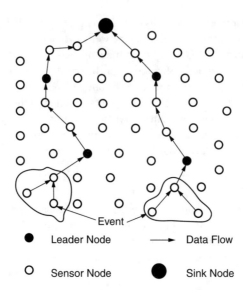

FIGURE 10.1   Mesh topology used for protocol evaluation.

10.4.1 Experimental Setup

In the simulation, 100 nodes are in a 1000- × 1000-m area, with one node at each of the positions of the 10 × 10 square grid. Figure 10.1 represents the mesh topology used for evaluating the protocol. The entire network is divided into five clusters. Two sensing areas are in the regions under clusters A and B. Sensor data packets are generated from these sensing areas at a uniform rate. The leader nodes in each of the clusters A and B collect these packets and send them to the leader nodes of clusters C and D, respectively, via intermediate sensor nodes. Leader nodes C and D forward these packets to the sink node in cluster E.

Each leader node selects the leader node that is geographically nearest to the sink for transmitting its received/sensed data. Leader–leader communication is accomplished through ordinary sensors. Reverse directional flooding is initiated when a leader node receives a sensor data packet indicating that at least *three* sensor nodes are close to the threshold *th*. A sender leader node sets *th* to the new βE_{min} obtained from the reverse flooding phase. The simulation is run for 900 sec with two leader–leader routing protocols: shortest path routing and the proposed team LEC protocol. These experiments are carried out on a simulation test bed that is an extension of Sensorsim [9].

10.4.2 Results and Analysis

It is assumed that before the network starts any activity, all ordinary sensor nodes have the same energy level; therefore, in the beginning, energy distribution is uniform across the network. When a network becomes active, the energy distribution across it gradually becomes nonuniform because nodes participating in a route inevitably consume more energy than other nodes do. A protocol that uses a fixed route until one node in the route is completely drained out of its energy ends up producing an energy distribution with high dispersion of energy levels. On the other hand, the proposed protocol tries to adapt to the dynamically changing energy distribution and gradually makes the initially uneven energy distribution uniform. Therefore, it is expected that the difference between the dispersion measures produced by this protocol and those produced by any protocol with fixed routing will increase with an increasing rate with time. This chapter compared the performance of this protocol with that of a protocol using a fixed shortest path for leader–leader communication.

Results of the simulation comparing performance of this protocol with that of shortest path routing reflect the outcome as expected. *Range* (the difference between the maximum and minimum value of a distribution) and *variance* were used as measures of dispersion of energy distribution to evaluate the protocol. Figure

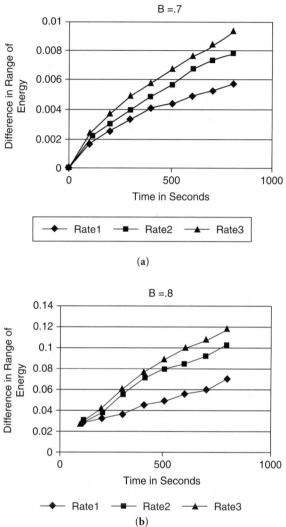

FIGURE 10.2 Difference in ranges of node energy distributions across the network over time under the two protocols for two different β values: (a) β = 0.7; (b) β = 0.8.

10.2(a) and (b) presents the difference in the ranges of node energy distributions across the network over time under the two protocols with two different values of β and three different traffic rates. In both figures, the difference of the ranges rises very sharply, indicating that this protocol yields a lower range of energy distribution compared to that produced by the fixed route protocol as time proceeds.

Moreover, with an increased traffic rate, this protocol produces a much better result compared to that of shortest path routing. This indicates that, with heavy traffic, the energy distribution across the network becomes more uneven in a fixed route protocol because the load is heavier on a particular route. In this case, frequent change of routing path is very useful in bringing uniformity to overall energy consumption. Note that routes changed more frequently with a higher value of β, performance of the proposed protocol is better when β = 0.8 than that when β = 0.7.

Figure 10.3(a) and (b) shows how the difference between the minimum energy level produced by this protocol and that by shortest path routing changes with time for two different β values, respectively. In both cases, the difference rises very sharply as time proceeds. With a higher value of β, the rise is sharper because change of route is accomplished more frequently; therefore, consumption of energy is more uniform under this protocol.

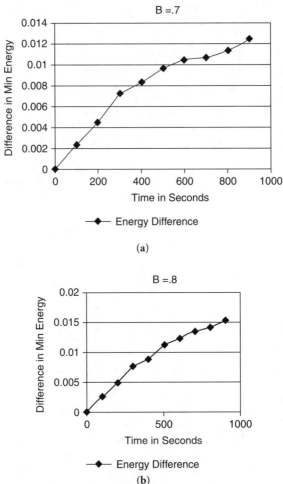

FIGURE 10.3 Difference between the minimum energy level produced by protocol and that by shortest path routing for two different β values: (a) β = 0.7; (b) β = 0.8.

Figure 10.4 represents the variance of residual energy distribution produced by this protocol with two different β values and that by the shortest path routing. The energy range metric does not measure the number of sensor nodes that are treated unfairly. The high variance of the shortest path routing indicates that a significant number of sensor nodes are treated unfairly, with network traffic concentrated at fewer nodes. This might expedite partition of the network due to energy depletion at critical nodes. With a higher value of β, the proposed protocol produces lower variance because of more frequent route changes.

Acknowledgments

This work was done in part with support from NSF under grants IIS-0312632 and IIS-0329738 and DARPA/AFRL under grant # F30602-02-1-0198.

References

1. I. Akyildiz, W. Su, Y. Sanakarasubramaniam and E. Cayirci, Wireless sensor networks: a survey, *Computer Networks J.*, 38(4), 393–422, 2002.

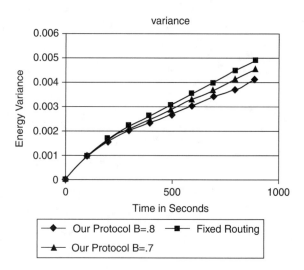

FIGURE 10.4 Variance of residual energy distribution produced by the protocol with two different β values and that by shortest path routing.

2. C. Intanagonwiwat, R. Govindan and D. Estrin, Directed diffusion: a scalable and robust communication paradigm for sensor networks, *Proc. 6th Ann. Int. Conf. Mobile Computing Networks (MobiCOM 2000)*, August 2000, Boston, MA.

3. M.J. Handy, M. Haase and D. Timmermann, Low energy adaptive clustering hierarchy with deterministic cluster-head selection, *4th IEEE Int. Conf. Mobile Wireless Commun. Networks*, Stockholm, 2002.

4. R.C. Shah and J.M. Rabaey, Energy aware routing for low energy ad hoc sensor networks, *Proc. IEEE WCNC'02*, March 2002.

5. Y. Yu, R. Govindan and D. Estrin, Geographical and energy aware routing: a recursive data dissemination protocol for wireless sensor networks, UCLA Computer Science Department technical report UCLA/CSD-TR-01-0023, May 2001.

6. B. Karp and H.T. Kung, GPSR: greedy perimeter stateless routing for wireless networks, *Proc. ACM/ IEEE MobiCom*, August 2000.

7. R. Kannan et al., Game-theoretic models for reliable path-length and energy-constrained routing with data aggregation in wireless sensor networks, to appear in *IEEE J. Selected Areas Commun.* (also as LSU Computer Science technical report LSU/CSC-TR03-05).

8. Max–min length-energy-constrained routing in sensor networks, in *Lecture Notes Computer Sci.* LNCS 2920, EWSN 2004, Berlin, Germany, January 19–21 2004. (submitted to IEEE JSAC).

9. S. Park, A. Savvides and M.B. Srivastava, SensorSim: a simulation framework for sensor networks, *Proc. MSWiM 2000*, Boston, MA, August 11, 2000.

10. D. Fudenberg and J. Tirole, *Game Theory*, MIT Press, 1991.

11. L. Zhou and Z. Haas, Securing ad-hoc networks, *IEEE Network*, 13(6), 24–30, 1999.

11

Localized Algorithms for Sensor Networks

Jessica Feng
*University of California at
Los Angeles*

Farinaz Koushanfar
University of California at Berkeley

Miodrag Potkonjak
*University of California at
Los Angeles*

11.1 Introduction

11.1.1 Motivation

Recently, wireless multihop networks (WMNs) have emerged as a promising architecture for realization of a various embedded distributed networked systems. WMNs can be used for a variety of tasks, including human communication and Internet-like data distribution. The most exciting application of wireless ad hoc networks is probably serving as the building platform for wireless sensor networks. In wireless sensor networks, each node is equipped with a certain amount of communication, computing, storage, sensing, and, in some scenarios, actuating resources. Wireless ad hoc sensor networks have the potential to bridge the gap between the Internet and the physical world. Numerous applications in the military environment as well as in personal and industrial tasks have been envisioned.

At the same time, wireless ad hoc sensor networks pose a number of new technological and optimization challenges. It is apparent that in order to address these challenges, sensor networks must operate in autonomous mode. In addition, in order to address low-energy, privacy, security, and scalability issues better, wireless sensor networks will require new types of algorithms that will use minimal amounts of communication. The goal of this chapter is to discuss the state of the art of algorithms commonly known as localized algorithms.

It is interesting to compare localized algorithms to other types of algorithms that have been excessively studied in computer science and related areas. In theoretical computer science and operational research, a great variety of algorithms has been developed for a wide range of combinatorial problems. These algorithms are developed under the following set of assumptions. The first is that constraints are on only two types of resources: storage and speed of computation. A number of models have been developed under this assumption, such as the Turing machine, Post's model, and the universal register machine. It has been demonstrated that these models are essentially equivalent. The inputs for the algorithm are specified at the beginning of its execution; run time and storage requirements serve as measurements of the quality of the solutions and algorithms. It is customary to consider algorithms that have run-time as polynomial functions with respect to the length of the input expressed in bits as efficient and the ones that require exponential time as inefficient.

On a more practical note, a number of paradigms that can be used to develop efficient algorithms have been identified, including divide and conquer; branch and bound; dynamic programming; and reduce and conquer. The key observation is that algorithms are designed and analyzed mainly based on how well they scale as the size of the input increases asymptotically. In addition, algorithms that guarantee optimal solutions and approaches guaranteeing that obtained solutions are within a certain vicinity of the optimal solution are widely studied (e.g., approximation algorithms), as well as algorithms that provide heuristic solutions when the problem is computationally intractable [3, 6, 7].

Although localized algorithms and even sensor networks have only been attracting research and development attention recently, already a wide literature and great variety of proposed approaches regarding the topic exist. It is already impossible to provide a comprehensive survey of all proposed algorithms for all wireless ad hoc sensor network tasks. The main objective in this chapter is to identify the most suitable abstractions and the most efficient techniques as foundations for developing localized algorithms. In addition, special emphasis is placed in summarizing how to analyze and evaluate localized algorithms. The goal is to cover all the most important developments as well as provide insights on why these algorithms are effective. In addition to presentation of already published results, several new algorithms that are optimal or superior to the published ones in terms of performance are proposed.

11.1.2   Chapter Organization

Section 11.2 summarizes all the proposed models, abstractions, and foundations for designing and analyzing localized algorithms in wireless sensor networks. In the next section, centralized algorithms that provide a comparison metric to localized algorithms are discussed. Section 11.4 presents several case studies for canonical problems in wireless sensor networks, as well as the existing algorithms, approaches and general paradigms. A number of widely applied analysis metrics and standards are presented in Section 11.5. In order to enable distributed localized algorithms, the different protocols in Section 11.6 can be applied in developing them; proposed techniques and algorithms for distributed localized algorithms are also discussed. Finally, Section 11.7 states some of the future conceptual, technological, and theoretical challenges related to localized algorithms.

11.2   Models and Abstractions

This section summarizes information about relevant models and abstractions required to specify and analyze localized algorithms. Much diversity is present among potential combinations of properties of models that can be used for this task. Many of them are interesting because they provide favorable trade-offs between their capability of capturing real-life sensor networks and their suitability for analysis and development of a variety of optimization techniques. Attention is focused on two groups: (1) those mainly related to widely used models in the literature; and (2) models favored by current and expected technology trends.

Currently, only static networks are considered when one studies models related to network topology. However, in the near future, a variety of models for mobile networks will appear. In order to ensure connectivity of all nodes, the standard assumption is that all nodes, when viewed at the graph level, form a single connected component. In addition, the edge between two nodes can be unidirectional or bidirectional. The first option is used when all nodes are equipped with identical radio transmitters and receivers. The second indicates situations in which node A can hear node B, but not vice versa. In addition, sometimes one or more nodes have special positions as gateways to the Internet or as base stations. The most important assumption about the network is related to the question of how much each node knows about the locations and connectivity of all other nodes.

The current standard assumption is that each node is only aware of its own neighborhood, i.e., nodes to which it can directly communicate. Sometimes this definition is enhanced to k-hop neighbors. In the future, schemes that explicitly state what is stored at each node will emerge. Essentially, as data structures play a crucial rule in the development of standard computer algorithms, data placement plays a crucial rule in localized algorithms. It is also important to note that as storage technology rapidly emerges, assuming that each node has only information about its own neighborhood is unrealistic. However, although information in static networks can be easily stored in each node, it would be expensive for each node to inform too many nodes about its status when the network is mobile or when an energy-saving procedure is conducted using sleeping mode.

Currently, it is most often assumed that nodes in the network are randomly deployed with uniform distribution in unit square areas. The assumption is justified in some scenarios, for example, when nodes are dropped from airplanes. However, it is obvious that new methodologies and approaches for WSNs with very different structural properties will emerge in order to address the needs of specific applications. In these networks, sensor placement will affect performance of localized algorithms in a very profound way.

Another aspect that is rarely discussed but crucially important is related to space topology and obstacles. For example, in environmental monitoring, simply ignoring trees and physical obstacles would inevitably result in incorrect conclusions. Finally, note that three-dimensional tasks are commonly significantly more difficult than two-dimensional tasks.

Currently, the standard assumption is that all nodes are equipped with identical transceivers and identical omnidirectional antennas. This assumption has the direct ramification that all two-communicating parties have the same transmission and reception strength. However, the communication range can be modeled in various ways depending on radios used. Four of the most intuitive options include:

- In the unit disk model, all nodes in the network have identical radio range.
- A generalization of the unit disk model is the arbitrary disk model, in which each node has an arbitrary radio range and is uniform along all directions. In this case, situations exist in which node A can hear node B, but node B cannot necessarily hear node A. Therefore, the arbitrary disk model requires directed graph for representation of the network connectivity.
- Another communication model relinquishes assumption of the uniformity of signal propagation along all directions and captures the statistical behavior of propagation signal as a probabilistic function of distance between the communicating node pair. Probability is different along different directions, but is a monotonically nonincreasing function along any given direction. Examples of the function that may be applied include the distance formula and the square of distance.
- Another option aims to incorporate complete arbitrariness in communication patterns. It assumes that communication between any two nodes, regardless of their positions, is established with a certain user-defined probability.

In addition to communication range, assumptions on the structure of transmitted data also play an important role in designing localized algorithms and evaluating their performance. The most widely adopted schemes are: (1) number of bits sent; (2) number of packets with no packet size restrictions; and (3) number of packets in which each packet has limited size.

The first option does not involve the concept of packet. Information is measured in terms of number of bits sent and received between two nodes that can communicate directly. The second scheme adopts the notion of packet, but packets are of a relatively large size relative to the information that must be sent so that they can be considered unlimited size packets. The last option imposes an upper limit of information that each packet can contain. Depending on the adopted communication models and the packet structure models, relative performances across different algorithms may significantly differ. Therefore, constructing algorithms most suited for the particular set-up so that they maximize the advantages of the assumptions is of great importance.

A number of energy consumption models exist. A specific example of an energy consumption model for wireless radio is given by Digitan. Assume 2 Mb/s 802.11; transmission takes 1.9 W of energy; reception takes roughly 1.5 W; idle/listening takes 0.75 W; and sleeping consumes only 0.025 W. The main observation is that unless the node is in the sleeping mode, no significant amount of energy can be conserved even if the node is in idle mode. The conclusion is simple and with strong ramifications: often it is more important to design localized algorithm that can be executed while a large percentage of nodes is in the sleeping mode.

Storage models can be categorized in two classes: direct and indirect storage. Direct storage implies that all the information each node stores is kept physically within the node. In indirect storage, data used by a node during execution of the localized algorithm are stored somewhere else — at some other node or possibly a separate gateway storage device. Therefore, this scheme requires an explicit step of referencing and communication in order to gain access to the information. Clearly, direct storage has advantages over indirect storage in terms of access time, flexibility, and communication cost. On the other hand, indirect storage can enable significantly better sharing of data as well as significant storage capacity enhancement.

Fault models are a well-studied topic and have been discussed comprehensively in VLSI and computer architecture literature. However, fault tolerance and therefore fault models have never been one of the dominating concerns and objectives for VLSI designs. The reason is that the properties of VLSI technology and design styles facilitate strong resiliency against faults naturally. However, wireless ad hoc sensor networks are vulnerable against faults (also equivalent attacks and data skewing) because of their wireless communication and localized mode of operation.

Furthermore, use of such networks also enhances the importance and the need for privacy and security. In addition, the observed physical world is full of obstacles that interfere with communication and sensing tasks. Sensor networks are often deployed in the physical world where the environment is complex or even hostile. For example, consider a habitat-monitoring sensor network deployed in a forest. Simply ignoring the existence of trees, plants, and other obstacles will lead to incorrect conclusions. Currently, fault tolerance is rarely addressed in sensor networks and the development of a fault-tolerant localized algorithm still must be addressed.

Sensing models capture sensitivity of a sensor as a function of parameters such as distances, properties of the environment, and position. For example, one can assume all sensors have only two sensitivity modes: detecting or not detecting an event. A widely used model for sensitivity is one in which the accuracy of sensing decreases according to a certain function of distance between the sensor and the target object. Linear and quadratic functions are often used.

11.3   Centralized Algorithm

This section discusses centralized algorithms for sensor networks. After the definition of centralized algorithms, their major advantages and disadvantages are briefly outlined. After that, several different scenarios in which centralized algorithms can be specified and analyzed are summarized. Special emphasis is placed on two phases: data collection and result dissemination. Several optimal centralized algorithms for common tasks in wireless sensor networks are presented.

Centralized algorithms in wireless ad hoc sensor network are procedures in which all information from all nodes in the network is first collected at a single, usually predefined, node. The problem is solved at this node and consequently the results of the optimization are disseminated to all nodes that requested this information. Therefore, three phases of centralized algorithms can be identified:

- Information collection in which readings of all sensors from all nodes are collected to a single computational point
- Optimization mechanism execution on that node
- Results of the optimization sent to all other nodes using multihop communication

One must study centralized algorithms for a given problem in which the primary goal is to develop the localized algorithm for several reasons. The first reason is that the centralized algorithm provides an upper bound of what is achievable with respect to the quality of the solution. At the same time, this algorithm also provides an upper bound of expected communication cost with respect to the corresponding localized algorithm. Note that both of the previous bounds are not actually guaranteed. For example, in the case of upper bound of the quality of the solution, if the problem is computationally intractable, it may happen that the localized algorithm "gets lucky" and produces a better solution than the centralized algorithm. In the case of communication cost, the centralized algorithm may get unlucky and some nodes are visited several times; therefore, energy consumption higher than the corresponding localized algorithm is the result.

There is a wide consensus that localized algorithms are the correct alternative for wireless ad hoc sensor networks. In a number of situations, centralized algorithms are obviously competitive if not better. For example, if the network is reasonably small and one must conduct several optimization problems at the same time, centralized algorithms are certainly attractive options to consider. Also, centralized algorithms are particularly well suited for a mapping problem in which each node must get a specific set of attributes.

It is important and interesting to consider relative advantages and disadvantages of centralized algorithms with respect to corresponding localized algorithms. In a number of aspects, centralized algorithms have significant advantages over localized algorithms. For example, the main logistic advantage is that optimization mechanisms do not need to be customized as in the case of localized algorithms. In addition, absolutely the same data collection and distribution algorithm and software can be applied to all problems. Furthermore, synthesis and analysis of centralized algorithms are significantly simpler conceptually and logistically than in the case of localized algorithms. For mapping problems, centralized algorithms are often competitive in terms of the communication cost. Finally, performance and cost of centralized algorithms most often have significantly lower variance in terms of quality of solution and communication cost than those of localized algorithms.

Nevertheless, localized algorithms have significant advantages in many situations that often greatly outweigh their limitations. For example, if size of the network increases, localized algorithms inevitably become the only realistic option. In particular, they show great advantages when search problems are addressed. Furthermore, localized algorithms provide strong advantages in terms of fault tolerance, security, and privacy. Finally, localized algorithms are much better suited for customization with respect to specific optimization mechanisms and communication models.

The advantages and disadvantages of centralized algorithms will be illustrated using several different abstractions and modeling scenarios. Three scenarios in which a centralized node is the Internet gateway that contains unlimited computation, storage, and energy supply resources will be considered. Note that, in this case, the centralized node has enough storage to contain all information regarding all nodes and their connectivity.

First consider a case in which the communication cost is measured in terms of transmitted data bits. This problem can be solved optimally. All that is required is that each node send its information using the shortest path to the centralized node. Dijkstra's algorithm can provide the solution in linear time in terms of number of edges in the graph. Notice that, because each node is sending information using the most sufficient route, the optimality of the algorithm is guaranteed.

In the second scenario, the communication cost is measured in terms of the total number of packets transmitted. In this case, the assumption that a packet has unlimited size is adopted; this is reasonable when the network is relatively small and the packet size limit is relatively large. In this case, the problem can also be optimally solved. The solution is based on the observation that each node must send its information at least once to some other node. Therefore, if the algorithm only requires each node to send its information once, the optimality is automatically achieved. The first step of the algorithm is to conduct breath-first search (BFS) in order to find the distance in terms of hops of each node from the centralized node. After that, each node in the network is scheduled to transmit its data or the data that it has received in a decreasing order according to its distance from the centralized node.

The third scenario is the situation in which the packet size is fixed to a certain amount and the goal is again to transmit the minimal number of packets. Unfortunately, the problem is now computationally intractable. Still, it can be solved optimally using integer linear programming (ILP)-based approaches. Note that, in many situations — particularly when the network is relatively small and sparse — this is attractive because it must be solved only once per lifetime for the network. The following variables are introduced:

$$X_{ij} = \begin{cases} m & \text{node } i \text{ sends } m \text{ bits to node } j \\ 0 & o/w \end{cases} \tag{11.1}$$

$$X_i = \begin{cases} l & \text{node } i \text{ sends } l \text{ outgoing bits} \\ 0 & o/w \end{cases} \tag{11.2}$$

$$Y_{ij} = \begin{cases} k & \text{node } i \text{ sends } k \text{ packets to node } j \\ 0 & o/w \end{cases} \tag{11.3}$$

There are two types of constraints. First, for each and every node, the outgoing number of bits that it sends out must equal the sum of the received bits plus the number of bits recorded. The second constraint ensures that the number of packets is sufficient to transfer the number of bits that need to be transmitted:

$$\left(\sum_{i=1}^{n} x_{ij} \right) + \left(R_i \right) = \forall i \tag{11.4}$$

$$y_{ij} > \frac{x_{ij}}{P} \tag{11.5}$$

where

R_i = number of bits that node i has recorded
P = packet size limit in terms of bits
n = total number of nodes in the network

The objective function is to minimize the number of total packets sent; therefore:

$$\text{min:} \sum_{i=1}^{n} \sum_{j=1}^{n} y_{ij} \tag{11.6}$$

Now consider the same three scenarios when there is no explicitly predefined centralized node. If the assumption is that each and every node is aware of the situation of the entire network, only minor modifications to the existing approaches would be sufficient. In the first scenario in which communication cost is measured in terms of bits, conduct the 1-to-n shortest path using Dijkstra's algorithm at each and every node, and select as the centralized node the one node with the smallest sum of shortest paths to all other nodes. In the case of the second scenario, in which the packet size is large enough to be considered unlimited size, all nodes have the same quality to be the centralized node; therefore, any arbitrary node can be served as the centralized node. In the case of the third scenario, in which the packet size is limited, one arbitrary node solves the system using the same ILP formulation with the assumption of a different centralized node, selects the node that provides the best objective function value when it is assumed to be the centralized node, and notifies this node to continue the procedure.

If the assumption is that each node only knows its limited neighborhood information, the problem becomes more complicated. In this case, the "spiral" algorithm [10] is proposed. Starting from an arbitrarily selected node, the goal is to minimize the number of times each node is visited in order to

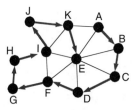

FIGURE 11.1 Example topology.

collect all the information in the network. The algorithm can be best understood in a geometric context. Consider the following illustrative example:

Figure 11.1 presents a network with 11 nodes; each is only aware of its own one-hop neighbors. Let node A be the arbitrary starting point; using the clockwise "sweeping" search technique, A finds the first occurrence of a nonvisited node, i.e., node B in this case. Therefore node A sends all its information to B. Now B applies the same technique to find the next first occurrence of a nonvisited node; this is node C. Node B forwards what node A has sent and node B's own data recorded to node C. This procedure continues until node E, which "sweeps" 360°. However, all the nodes encountered have been visited, so node E concludes that it has all the information in the network and announces that it is the centralized node.

Once all the information is present at the centralized node, it can apply various optimization techniques to obtain solutions. Focus on the last phase of the centralized algorithm — solution dissemination. The problem is equivalent to the broadcasting problem, which can be again addressed using ILP. Define the following variables:

$$X_i = \begin{cases} 1 & node\ i\ broadcasts \\ 0 & o/w \end{cases} \tag{11.7}$$

$$X_{ij} = \begin{cases} 1 & node\ i\ sends\ message\ to\ j \\ 0 & o/w \end{cases} \tag{11.8}$$

Using these specified variables, the following three constraints are enforced. First, each node must receive the information from some other node in the network. The second type of constraint ensures that only nodes within communication range of each other can communicate. The third type of constraint ensures that the broadcasting node is only charged once no matter how many nodes have received messages from it.

$$\sum_{\substack{i=1 \\ i \neq j}}^{n} x_{ij} \geq 1 \qquad j = 1,...,n \tag{11.9}$$

where n = total number of nodes in the network.

$$x_{ij} \qquad if\ E_{ij} \neq 1 \tag{11.10}$$

$$x_{ij} \leq x_i \qquad i = 1,...n; \quad j = 1,...n; \quad i \neq j \tag{11.11}$$

The objective is again to minimize the number of packets sent, i.e., minimize the number of nodes that broadcast:

$$\text{min:} \sum_{i=1}^{n} x_i \qquad\qquad (11.12)$$

11.4   Case Studies

11.4.1   Energy Management and Topology Maintenance

A number of alternative power minimization methods act above the MAC layer powering off redundant nodes' radios in order to expand the battery lifetimes. For example, AFECA [19] trades off energy consumption and the quality of the message delivery services based on the application requirements. GAF [20] is another power-saving scheme that saves energy by powering off the redundant nodes. GAF identifies the redundant nodes by using the geographic location and a conservative estimate of the radio ranges. It superimposes a virtual grid proportional to the communication radius of the nodes onto the network. Because the nodes in one grid are equal from the routing perspective, the radios of the redundant nodes within a grid can be turned off. The nodes awake within a grid rotate to balance their energy.

One of the main advantages of GAF is that it is completely static and localized. All nodes are capable of estimating virtual grids and determining equivalent nodes. In addition to saving 40 to 60% of the energy compared to an unmodified ad hoc routing protocol, GAF also suggests that network lifetime increases proportionally to node density. On the other hand, a significant performance bottleneck can be easily created by grids that contain very limited number of nodes. Moreover, sometimes it is acceptable to let all nodes in some grids sleep (e.g., the boundary nodes) in order to reduce energy further. However, this situation cannot be recognized by GAF.

SPAN is a power-saving, distributed, randomized coordination approach [1] that preserves connectivity in wireless networks. The work presented in Koushanfar and colleagues [9] has proved the necessary and sufficient conditions for putting the radios in the sleep mode, while still guaranteeing connectivity. A major advantage of this scheme is that all the decisions are made locally and individually. Therefore, it is much more robust, flexible, and scalable than the centralized schemes. In addition, according to the condition of the network, coordinator nodes are adjusted and re-elected locally as well. However, SPAN shares some similar limitations with GAF, in particular with respect to energy savings. For example, in some situations not all coordinator nodes need to be awake.

There are also a number of research efforts that trade off between latency and energy consumption. The power management approach presented in Kravets and Krishnan [12] selectively chooses short periods of time to suspend and shut down the communication unit; they queue the data before suspending the communication. STEM is a power-saving strategy [17] that does not try to preserve the capacity of the network. STEM works by putting an increasing number of nodes into sleep mode, and then encountering the latency to set up a multihop path. Nodes in STEM must have an extra low power radio (paging channel) that does not go into sleeping mode and constantly monitors the network to wake up the node in case of an interesting event.

11.4.2   (MI)^2

In traditional computer science, backbones for designing efficient algorithms are optimization paradigms such as branch-and-bound, dynamic programming, divide-and-conquer, and iterative improvement. This section introduces the maximally informed maximally informing (MI)^2 paradigm — the first systematic approach for the design of localized algorithms. In order to make the presentation self-contained, key assumptions are first summarized and typical sensor network optimization problems that will serve as illustrative examples briefly described. After that, an explanation is offered on how to apply the (MI)^2 strategy in a systematic way during each of the four phases of a localized algorithm: information gathering, system structuring, optimization mechanism, and result dissemination. Key insights and key trade-offs in designing localized algorithms are described. The realization of such algorithms on a number of typical sensor network tasks, such as routing and minimum spanning tree, is illustrated.

Given a network, assume that each node has minimal state information about the network and is only aware of nodes within its communication range. This is so because: (1) it is necessary to minimize storage requirements at each node; (2) nodes go to sleeping mode from time to time in order to minimize the energy consumption [2, 16]; and (3) updating the routing tables might not be possible as a result of nodes' high mobility.

The goal of the shortest path problem is to find a path between S and D such that the path has the smallest cardinality (i.e., the smallest number of nodes on the path). The MST problem asks to find the minimum spanning tree for a subset of nodes in the network. The connected dominating set problem addresses selecting a subset of nodes of minimal cardinality in such a way that each node is in the subset or has a neighbor in the subset. The importance of the selected problems for wireless ad hoc networks is self-evident. For example, the connected dominating set ensures that information can be efficiently collected or distributed from the nodes in the dominating set to all other nodes [18].

Although previous research in this area has implicitly specified the four phases in the design of localized algorithms, the phases are explicitly identified and formalized here for the first time. More importantly, the novelty of this approach is that insights and systematic generic methods to leverage the $(MI)^2$ paradigm have been developed in each step. This results in efficient localized algorithms on a great variety of problems.

11.4.2.1 Phase 1: Information Gathering

The information gathering (IG) phase is where the inputs to the procedure are prepared. If the information from multiple nodes is needed, routing between the nodes and the order in which nodes are visited and information is gathered will have a large effect on the amount of energy consumed. According to the maximally informing paradigm, each step of the IG phase must select the next node to be visited or contacted in such a way that the maximal amount of relevant information required for the application of the optimization mechanisms is acquired. The maximally informing principle can be realized in several ways, depending on the considered scope and objective function of the optimization problem. When considering the scope, one can take a greedy local view in which one considers which nodes can be contacted in a few hops if a particular node is visited next.

When considering the objective function, one can contact a node that will expose the largest number of constraints itself, or contact a node that has neighbor nodes that will reveal the largest number of constraints. For example, one alternative is to select a node that is likely to have many unvisited neighbors as the next node. In this case, the amount of obtained information is maximized. Another alternative is to visit a node that has a large unexplored area within its communication range with a high likelihood of containing nodes relevant for optimization.

For example, in shortest path routing, one can always contact the node closest to the destination node in a greedy way. An alternative is to contact the node with the largest area in its communication range, with a large percentage of points that are closer to the destination.

The final important observation related to the IG phase is that, in certain situations, visiting some nodes is perhaps more important than visiting others. One such situation is when the goal is to find the connected dominating set for all nodes in a geographic region. In this situation, it is crucial to visit all nodes on the outer perimeter of the network because their information could guarantee that all of the relevant nodes are considered. Therefore, in this situation, the $(MI)^2$ paradigm indicates that these nodes should be visited first.

11.4.2.2 Phase 2: System Structuring

Every node in the system has some amount of processing capability. However, not all of the system nodes need to compute the optimization procedure all the time. In the system-structuring phase, the decision about when and where to conduct optimization mechanism computations is made.

According to the $(MI)^2$ paradigm, two principles for selection of computation centers are followed. The first is to assemble enough information initially to conduct at least part of the computation meaningfully as soon as possible. This point is particularly well illustrated on MRA [9] and exposure tasks [13]. The second principle is always to conduct computations at the boundary of an already visited region in order to reduce the requirements for obtaining additional information. This point is clearly illustrated with the exposure task.

Finally, note that different optimization mechanisms dictate different system structuring phases. In some tasks, such as MRA, the local information is sufficient to guarantee the optimum solution. However, in computationally intractable optimization problems in which the interaction between all of the nodes in the system defines the output, the quality of the solution may be seriously hampered using only localized scopes. In such situations, it is necessary to obtain information about a large neighborhood for each node before the optimization mechanism is started.

11.4.2.3 Phase 3: Optimization Mechanism

Once the needed input is at a computation center, the optimization procedure is executed. The separation-of-concerns principle suggests that the phases should be as independent as possible. However, in the majority of problems, strong interdependence exists between the information-gathering phase and the optimization mechanism (OM) phase because, based upon the specific needs for executing an optimization mechanism, the relevant information must be acquired.

The first observation is that constructive and deterministic algorithms are strongly preferred to iterative improvement and probabilistic algorithms. This is because the former algorithms require only one pass through all inputs, but the latter require multiple passes. For example, for the MST problem, Prim's algorithm is much better suited for implementation as a localized algorithm than Kruskal's algorithms. This is the case because Prim's MST algorithm starts from an arbitrary node and at each step selects the shortest edge incident to one of the nodes already visited and does not form a cycle with the edges in the existing partial MST. This edge is then added to the partially built MST. Therefore, the algorithm uses only information about nodes that are already visited and their neighbors. On the other hand, Kruskal's algorithm requires one to consider all edges in the graph at each step and select the globally shortest edge. Therefore, before starting the execution of Kruskal's algorithm, it is necessary to obtain information about the whole graph.

The $(MC)^2$ optimization paradigm is well suited for use in conjunction with the $(MI)^2$ paradigm. In order to gain maximal benefit from the merged $(MC)^2(MI)^2$ paradigm, it is often advantageous to consider variants of $(MC)^2$ that only consider nodes adjacent to already explored nodes. This must be done in such a way that communication requirements are reduced. Other optimization paradigms naturally well suited for design of localized algorithms and, in particular with the $(MI)^2$ paradigm, are branch and bound and dynamic programming-based algorithms. Finally, note that in some cases, such as exposure calculations, one can directly use the available optimization mechanism. In others, such as the MRA problem, in order to design an efficient localized algorithm, one must develop a new optimization mechanism and, therefore, a new centralized algorithm that operates locally and with the partial information.

11.4.2.4 Phase 4: Information Dissemination

The information dissemination phase is the step in which the output of the optimization procedure is sent to the nodes requiring that information. The maximally informed paradigm states that one should disseminate information about the output of the optimization node to a particular node while close to that node. In the ideal case of balanced optimization and information distribution phases, all information that some node requires should be sent when visiting the last of its neighbors.

11.4.3 Solving ILP Problems by $(Mi)^2$-Based Paradigm

To demonstrate the wide application range of a paradigm for designing localized algorithms, apply it to a set of specific problems that can be specified and solved using a particular optimization solving strategy.

This subsection presents an $(MI)^2$-based approach for solving an instance of a problem specified as an integer linear program. ILP is a widely used procedure for specifying and solving combinatorial optimization problems. ILP formulations are readily available for a large variety of combinatorial optimization problems, or they are easy to develop [14]. In particular, in a special case of ILP, called 0-1 ILP, all variables must be assigned to one of two binary values [14]. For example, all problems discussed in this chapter can be easily specified and solved using a 0-1 ILP formulation.

ILP formulation has three different components: variables, objective function, and constraints. Variables can take only integer values; the objective function and constraints must be linear. Note that, if the requirement that variable must be integers is removed, ILP reduces to a linear program (LP) that also has a wide range of applications [15]. An ILP defined over a set of variables x_i has the following standard form:

$$\text{Max } E^T.\mathbf{X} \tag{11.13}$$

$$\text{such that: } A^T.\mathbf{X} \ B, C^T.\mathbf{X} = D \tag{11.14}$$

where A, B, C, D, and E are matrices composed of real constants, and $\mathbf{X}$ is vector consisting of variables x_i. The first clause is the objective function (OF), while the equations on the second line are the constraints.

Assume that each node has information about one or more coefficients from matrices A, B, C, D, and E. Furthermore, the node has a list of its neighbors and a list of information of each neighbor, but does not necessarily have all the information that each neighbor has. The reason for this assumption is that, for many optimization parameters (such as energy level, sleep state, occupancy of buffers), collected sensor information is transmitted only on demand in a sensor network in order to reduce power consumption. Finally, each node must be informed about the value of all variables x_i important to it.

The $(MI)^2$-based approach for locally solving an ILP instance is based on the following observation and intuition: a particular value can be assigned to a particular variable x_i, only after information is obtained about all constraints that contain x_i. Furthermore, it is advantageous first to resolve variables that are components of the most difficult (strict) constraints. Also, it is important that at the time of assigning a particular value to a variable, as much information as possible is available about all constraints that contain the variables contained by the constraint under consideration. In order to maximize the objective function, it is important to assign high values to variables with high coefficients and to keep estimating the highest possible value of the OF in view of the already observed constraints.

One can envision two approaches with respect to the relationship between the OF and constraints: optimistic (in which it is preferable to maximize the objective function at potential danger that later some constraints will become unable to be satisfied), and pessimistic (in which constraints are favored at the expense of the OF). The $(MI)^2$-based localized ILP procedure is summarized using the pseudocode presented in Figure 11.2.

The IF is weighted sum of resolving power of the information available at the node and resolving power of its neighbors. The weighs of neighbors to a node are scaled by the average number of neighbors

---

Procedure (*Localized $(MI)^2$-based ILP Procedure*)

Initialization;
while (*termination criteria* ==No) {
        Contact a neighbor that has highest information function (IF);
if (there are neighbors that do not have unvisited neighbors) {
        execute the optimization mechanism and communicate
        assigned values of the assigned variables to them } }

FIGURE 11.2 Pseudocode for $(MI)^2$-based localized ILP procedure.

from already visited nodes. Resolving power is proportional to reduction in information uncertainty, according to the classical information theoretical definition.

The optimization mechanism used is based on the maximally constrained, minimally constraining principle. Essentially, one tries to assign each variable in such a way that it resolves a maximal number of constraints or increases the chance that they are later satisfied. The optimization function is treated as a constraint that is dynamically updated. The initial value is provided by a simple probabilistic analysis and consequently the value is updated by extrapolating the values obtained from the already visited nodes.

11.4.4 GPSR

Routing is one of the fundamental tasks in wireless networks. Although one can envision a number of different types of routing, the focus here is on a case in which a single message must be sent from a node to another node. Only one localized routing algorithm will be considered so that it can be described and analyzed in sufficient detail.

Karp and Kung [8] have developed a stateless routing protocol for wireless networks: greedy perimeter stateless routing (GPSR). The development of GPSR is based on two main assumptions. First, it assumes that each node (router) in the network is aware of its geographic location and the geographic locations of all its direct (one-hop) neighbors. Second, it assumes that the geographic location of the destination is also known. GPSR abandons traditional routing concepts that require continual distribution of the current map of the entire network's topology to all nodes. The packet forwarding decisions are made based only on the positions and knowledge of local nodes and the final destination location. More specifically, each node considers the locations of all neighbors, and makes a greedy decision to forward the data packet to the node closest to the destination. Therefore, GPSR is stateless in the sense that it does not keep additional information about the rest of the network beyond its neighborhood. As a consequence, GPSR scales better than traditional routing protocols and is much more adaptive to mobility.

GPSR protocol has two phases: greedy forwarding and perimeter forwarding. Greedy forwarding refers to the phase in which a series of nodes follow the same rule and each node makes a greedy decision of forwarding the data packet to the one neighbor that the current node believes is the closest to the destination. However, greedy forwarding would fail in a situation in which a node is the local minimum in terms of its geographic distance to the destination, i.e., when all its neighbors have longer distances to the destination than it does. In this case, control is switched to perimeter forwarding mode from greedy forwarding in order to escape the deadlock. Perimeter forwarding essentially follows the right-hand rule, which seeks to find an alternative route around and eventually converges to the destination.

In addition to being stateless and having exceptional scalability, GPSR has a number of other noble properties. It is efficient in the sense that it often selects the optimal or near-optimal path when the network is dense. It is also conceptually (and from implementation point of view) very simple and clean. However, it has a number of limitations. For example, if the network is not very dense, it is easy to show that the greedy approach is not the best choice because the scope of the problem considered is limited with respect to available information. In addition, there is no guarantee that GPSR will eventually converge to the destination. Situations exist in which forwarding phase and perimeter phase oscillate within a set of nodes and never converge on the correct destination. It is also difficult, if not impossible, to see how to generalize the approach when additional information is available to a three-dimensional case, or in the presence of obstacles.

11.5 Analysis

Creation of algorithms has two interdependent phases: synthesis and analysis. Although synthesis of localized algorithms is widely considered a difficult and demanding task, analysis often does not receive the proper attention and treatment. In this section, the most important issues related to analysis of localized algorithms are discussed.

Analysis of localized algorithms can be defined as a process of characterizing the effectiveness of a proposed localized algorithm for a given problem. It is a complex and often cumbersome task for several reasons. First, it is not easy to identify which properties of the algorithms are interesting and important. Even when these properties are identified, it is often unclear how to define each of them exactly. In addition, it is often difficult to calculate or measure these properties. For example, some are associated with solving computationally intractable problems.

The next layer of complexity comes from a need to consider more than one property simultaneously. Furthermore, it is not clear *a priori* what should be the representative and realistic properties of instances. Finally, one can consider localized algorithms as generalizations of on-line algorithms in which the designer has an impact on information that will be obtained next. Therefore, unpredictability often results in randomness of characteristics of a particular algorithm.

The primary goal of localized algorithms is to minimize the amount of energy spent on communication. This does not necessarily mean minimization of the number of packets. The current technology indicates that the most effective way of saving energy is through placing the radios of as many nodes as possible into sleeping mode. Also, note that in future applications, energy minimization will not be necessarily equivalent in the first approximation the minimization of energy devoted to communication. Depending on the technology, and even more on the targeted applications, computation or some other components may have the dominant role.

The primary constraint is to achieve the user-requested level of optimality and/or accuracy. Because of complex error propagation through the sensor fusion phase, it is sometimes difficult to select the most appropriate definition of accuracy.

Historically, the performance of algorithms has been evaluated as the size of their input asymptotically increases. Also, in traditional networking research, one of the key issues is scaling the protocols as the size of the network increases. Although many wireless sensor networks will be of limited size, scaling localized algorithms is already widely studied. A better way to evaluate localized algorithms for limited-sized networks is probably the development of benchmarks. Unfortunately, of the very few benchmarks available at present, all are synthetic.

In addition to these three metrics, amenability to provide fault tolerance, satisfy real-time constraints (such as throughput and latency), maintain privacy and security, and facilitate mobility will also be of prime importance for evaluation of localized algorithms.

One can envision many ways to combine two or more metrics. For example, in operation research literature, it is common to derive a set of solutions that form a Pareto optimal curve. In the computer science literature, it is more common to take one metric as the optimization goal and others as constraints.

11.6 Protocols and Distributed Localized Algorithms

This section briefly discusses the distributed localized algorithms in which more than one thread of computation is executed at the same time. Distributed localized algorithms have a number of advantages in terms of their ability to respond faster to changes in the environment and the network, fault tolerance, and their resiliency against security attacks. First the desiderata for protocols that govern the execution of distributed localized algorithms are stated. After that, one generic approach is presented for development of protocols for distributed localized algorithms [9].

Proper computation and synchronization strategy should have the following characteristics:

- *Concurrency.* The computation (decision making) should take place at as many places in the network as possible. In particular, nodes should be constantly updating their resources to cope with the dynamics in the network.
- *Synchronization (avoiding deadlocks).* The computing nodes should not have a conflict on the resources they use. For example, assume that a node $v1$ finds a node $v2$ redundant in terms of a specific functionality. At the same time, $v2$ also finds $v1$ redundant. If $v1$ and $v2$ decide to go to

sleep (using each other as a back up), a deadlock will occur. A good synchronization strategy must avoid deadlock situations like this.

- *Overhead.* The computation and synchronization strategy should add an overhead as low as possible to the network, especially in terms of its power consumption and communication overhead.
- *Latency.* Higher latency in putting a node into sleeping mode implies more idle energy consumption. Also, nodes should be updated for changes in the network to adapt to the network dynamics.
- *Fault tolerance.* Fault is inevitable in sensor networks. The computation and synchronization strategy should be designed so that the faults in any number of nodes cannot corrupt its functionality.

Koushanfar et al [9] have developed an approach termed "distributed token mechanism" that attempts to fulfill the stated requirements. A token indicates that the node has control of the local flow of the sleeping procedure. At each point of time, more than one token can be present in the network to comply with the concurrency requirements. A token is generated by an awakened node that needs to check the eligibility of the nodes within its local scope to enter the sleep state. The token is eliminated as soon as it examines the functionality of its local scope of the network and selects the nodes for sleeping. The node with the token locks its local area of consideration so that no other nodes use the same resources and the nodes acting on the mutual resources are synchronized. To lock a node means to consider it only for one token at each point of time.

The localized and distributed nature of the token generation makes it very tolerant to faults at the individual nodes. The pseudocode for the distributed token mechanism procedure is shown in Figure 11.3; a node v_i that is not already locked by any other nodes considers running the sleeping procedure (steps 1 through 4) and therefore generates a token. A node that has slept before generates the token at a random time r_i within the interval ($0 < r_i < r$max) (steps 5 through 9); a node that has already changed its state into sleep at least once generates a token as soon as it wakes up (steps 10 and 11). A node with the token locks all of the unlocked nodes within its local scope of consideration (step 12). This node then runs the sleeping procedure, which decides which of the locked nodes can enter the sleep state (step 13) and for how long (step 14). The token node then announces the decision to its neighborhood (step 15) and unlocks the locked nodes (step 16).

The random initiation time (r_i) assigned to each node in the beginning of the procedure serves the purpose of avoiding simultaneous requests on the use of mutual resources. Because the sleep intervals are assigned independently to nodes depending on the power and topology of the neighborhood, the

```
Procedure Distributed Token mechanism

1.   at ∀ node vᵢ,
2.   {
3.       while (node vᵢ is not locked by another node)
4.       {
5.         if (never have slept before)
6.         {
7.            set a random initiation time rᵢ (0 < rᵢ < rₘₐₓ);
8.            generate a token at the time rᵢ;
9.         }
10.      else {
11.            generate a token as soon as vᵢ wakes up;
12.            lock all the unlocked nodes in the vᵢ's scope;
13.            select the best node to sleep;
14.            select the sleep interval for the sleeping node;
15.            announce the decision in the neighborhood;
16.            unlock the locked nodes;
17.        }
18.    }
```

FIGURE 11.3   Pseudocode for the distributed token mechanism procedure.

wake-up times are different. Therefore, after a node wakes up, it can immediately start another round of sleeping strategy without having too many locked nodes in its neighborhood. It is reasonable not to be concerned about the collisions in the network because they rely on the network's MAC layer to resolve any such conflicts.

11.7 Pending Challenges

This section outlines some of the potential trends for developing localized algorithms. It is always dangerous to make predictions, in particular when the topic is broad and application dependent; nevertheless, one can identify some major trends. Future research directions are classified into two broad categories related to: (1) the conceptual novelties for developing localized algorithms; and (2) optimization and algorithmic techniques. Due to space limitations, many important directions, such as interaction of localized algorithms with privacy and security; mobility; fault tolerance; applications within real-time systems; and use for actuator-based system, are omitted.

It is well known that mandatory prerequisites for developing high-quality algorithms are sound theoretical foundations. In some cases, one can develop such foundations, for example, PRAM, URM and the Von Neumann models of computation. When it is difficult to define a single widely applicable model, such as in parallel computing, progress is much slower. Currently, several models have been proposed for wireless ad hoc networks, including that of Zonoozi and Dassanayake [21]. However, it seems that the completely random nature of these models makes them of relatively limited practical relevance. Several other fields have also developed theoretical models. For example, in VLSI computations, the standard model is the one that assumes planarity and finite feature size of transistors and interconnects. The development of sound foundations for wireless sensor networks is a complex and difficult task because one must model at least four aspects of the systems: computation, communication, storage, and sensing.

Future algorithmic techniques can be naturally classified into two groups: one is related to design and the other is related to the analysis of localized algorithms. Design-related issues include the development of new paradigms that will facilitate systematic creation of localized algorithms, in particularly data collection and dissemination. An example of this is the maximally informing and maximally informed paradigm [11].

Currently, although a number of localized algorithms have been published, relatively little is known about their optimality in terms of quality of solution and expected energy cost. Several approaches have been proposed for this purpose, including the development of low bounds and probabilistic analysis. This trend will continue and will include new hard bound techniques as well as statistical guarantees.

Obviously, a strong correlation exists between how nodes are deployed and performances of localized algorithms. It is easy to see that different localized algorithms are best suited for different wireless sensor network organizations. Interestingly, this topic has not been addressed. In particular, sensor networks with regular structure such as grid can facilitate the development of fast and efficient localized algorithms. Another important issue with respect to localized algorithms for sensor networks is the development of optimization mechanisms that are resilient against unavoidable errors in sensor measurements. Finally, there will be a particular need to develop comprehensive approaches that combine continuous, discrete, and statistical techniques in order to obtain efficient localized algorithms. An example of this is exposure coverage [13].

Another side of the coin for localized algorithm development is the analysis of localized algorithms. Soon many activities will be conducted to define and develop scalable algorithms that are scalable not only with respect to the size of the network, but also with respect to the intensity of errors and the quality of solutions. Localized algorithms are, in a sense, the generalization of the concept of on-line algorithms in which one can decide which piece of information to obtain next. Competitive analysis of on-line algorithms has been a widely studied topic; it will be important for localized algorithms as well.

From a practical point of view, the most urgent issue is to develop standard benchmark examples that can properly capture the properties of real-life applications. Once the benchmarks are available, it would be important to analyze localized algorithms using statistical and perturbation analysis [4, 5].

A new network (distributed systems) architecture will appear, and it will be well suited for specific classes of tasks and applications. In addition, there is an urgent need for rapid prototyping and simulation platforms on which performances of localized algorithm can be accurately observed and quantified. Another important research direction is the development of design patterns for common localized algorithms. Design patterns changed the way in which software development is conducted and it will have a high impact in sensor networks.

Acknowledgment

This material is based upon work supported in part by the National Science Foundation under Grant No. ANI-0085773 and NSF CENS Grant.

References

1. Chen, B. et al. Span: an energy-efficient coordination algorithm for topology maintenance in ad hoc wireless networks, *Int. Conf. Mobile Computing Networking (MOBICOM)*, 85, 2001.
2. Estrin D. et al. Next century challenges: scalable coordination in sensor networks, *Int. Conf. Mobile Computing Networking (MOBICOM)*, 263, 1999.
3. Goemans, M.X. and Williamson, D.P. Improved approximation algorithms for maximum cut and satisfiability problems using semidefinite programming, *J. Assoc. Computing Machinery*, 42, 1115, 1995.
4. Grossglauser, M., and Tse, D.N.C. Mobility increases the capacity of ad hoc wireless networks, *IEEE/ACM Trans. Networking*, 10, 477, 2002.
5. Grossglauser, M., and Vetterli, M. Locating nodes with ease: mobility diffusion of last encounters in ad hoc networks, *Annu. Joint Conf. IEEE Computer Commun. Soc. (INFOCOM)*, 1954, 2003.
6. Hochbaum, D., Ed. *Approximation Algorithms for NP-hard Problems*, PWS Publishing Company, 1997.
7. Johnson, D. Approximation algorithms for combinatorial problems, *J. Computer Syst. Sci.*, 9, 256, 1974.
8. Karp, B. and Kung, H.T. GPSR: greedy perimeter stateless routing for wireless networks, *Int. Conf. Mobile Computing Networking (MOBICOM)*, 243, 2000.
9. Koushanfar, F. et al. Low power coordination in wireless ad hoc networks, *Int. Symp. Low Power Electron. Design*, 475, 2002.
10. Koushanfar, F. et al. Algorithms for resource discovery in wireless networks, unpublished manuscript, 2003.
11. Koushanfar, F. et al. Maximally-informing and maximally-informed algorithms for wireless networks, unpublished manuscript, 2003.
12. Kravets, P. and Krishnan, P. Application-driven power management for mobile communication, *Wireless Networks*, 6, 263, 2000.
13. Meguerdichian, S. et al. Exposure in wireless ad hoc sensor networks, *Int. Conf. Mobile Computing Networking (MOBICOM)*, 139, 2001.
14. Nemhauser, G.I. and Wolsey, L.A. *Integer and Combinatorial Optimization*, John Wiley & Sons, 1988.
15. Papadimitriou, C. and Steiglitz, K. *Combinatorial Optimization: Algorithms and Complexity*, Prentice Hall, Englewood Cliffs, NY, 1982.
16. Rabaey J.M. et al. PicoRadio supports ad hoc ultra-low power wireless networking, *Computer Mag.*, 42, 2000.

17. Schurgers, C., Tsiatsis, V., and Srivastava, M. STEM: topology management for energy-efficient sensor networks, *IEEE Aerospace Conf.*, 78, 2002.

18. Stojmenovic, I., Seddigh, M., and Zunic, J. Dominating sets and neighbor elimination based broadcasting algorithms in wireless networks, *IEEE Trans. Parallel Distributed Syst.*, 13, 14, 2002.

19. Xu, Y., Heidemann, J., and Estrin, D. Adaptive energy-conserving routing for multihop ad hoc networks, technical report *527*, USC/Information Science Institute, October, 2000.

20. Xu, Y., Heidemann, J., and Estrin, D. Geography-informed energy conservation for ad hoc routing, *Int. Conf. Mobile Computing Networking (MOBICOM)*, 70, 2001.

21. Zonoozi, M. and Dassanayake, P. User mobility modeling and characterization of mobility patterns, *IEEE J. Selected Areas Commun.*, 15, 1239, 1997.

Index

F